# Charcoal Analysis:
# New Analytical Tools and
# Methods for Archaeology

## Papers from the Table-Ronde held in Basel 2004

Edited by

## Alexa Dufraisse

BAR International Series 1483
2006

Published in 2016 by
BAR Publishing, Oxford

BAR International Series 1483

*Charcoal Analysis: New Analytical Tools and Methods for Archaeology*

ISBN 978 1 84171 918 4

BAR Publishing is the trading name of British Archaeological Reports (Oxford) Ltd.
British Archaeological Reports was first incorporated in 1974 to publish the BAR
Series, International and British. In 1992 Hadrian Books Ltd became part of the BAR
group. This volume was originally published by Archaeopress in conjunction with
British Archaeological Reports (Oxford) Ltd / Hadrian Books Ltd, the Series principal
publisher, in 2006. This present volume is published by BAR Publishing, 2016.

Printed in England

# BAR
PUBLISHING

BAR titles are available from:

BAR Publishing
122 Banbury Rd, Oxford, OX2 7BP, UK
EMAIL   info@barpublishing.com
PHONE   +44 (0)1865 310431
FAX     +44 (0)1865 316916
www.barpublishing.com

# AKNOWLEDGEMENTS

First, I wish to thank Stefanie Jacomet for her encouragement and helpful comments during the preparation of this meeting, and Fritz Schweingruber for his participation and foreword.
I would like to address my thanks to all the participants of the Round Table, hoping that this meeting will necessarily initiate new workshops.
It is also a pleasure for me to express my gratitude to the members of the IPSA, especially Jorg Schibler, who contributed to the success of this meeting.
I would also like to thank Danièle Martinoli and Petra Zibulki who dedicated their time and efforts to this meeting, and to thank them also, along with Britta Pollman, for their help in translating some abstracts into German. Thanks are also due to Christophe Croutsch who collaborated on the paging of this manuscript and to Isabelle Jouffroy for the references review.

Alexa Dufraisse

Besançon, the 15th of November 2005

# CONTENTS

# FOREWORD

## Charcoal

Few million years ago "the first man" learned to use fire. Charcoals are the first witnesses of the clad Homo. Few thousands years ago people learned to manage fire by producing charcoals. That technique opened the production of metal and finally to produce trains and cars. Today engineers analyse physical and technical features of charcoal and archaeologists try to reconstruct prehistoric vegetation and human activities. The global presence of charcoal is a tremendous source of information because is resistant against biological degradation, remains at the original place and cam be dated by the radiocarbon method.

The charcoal meeting at Basel covered a small, but an important aspect of the scientific potential of the charcoal analysis, the anthracology. The full scientific importance can be evaluated by few observations. Most important is the stratigraphic position. If that is clear results from charcoal analysis allow comparisons with archaeological findings, sedimentological qualities. In the past charcoal analysis concentrated on species identification. But the fragmentation, the form of particles and their consistence, the intra-tree provenance, the fungi content and radiocarbon and dendrochronological dates are important "criminalistic" features. All the characteristics allow the reconstruction of human activities all over the world without any restriction. Man lived in caves at wet places and dry land in hot deserts in cold steppes north of timberline at sea shores and alpine meadows. Mans witnesses, the charcoals, are everywhere.

The charcoal meeting also clearly demonstrated the handicaps. Charcoal identification is geographically restricted because we don't have wood identification atlases for many regions of the world. We miss bark, twig and root atlases from living trees, therefore bark identification is hardly possible. Many knowledge's about the physical characteristics are known in technical sciences but the haven't be acknowledged enough in the science of anthrocology. And last but not least: charcoal analysis never stands alone. It is always embedded in an array of results of other scientific methods.

Fritz H. Schweingruber

Birmensdorf, the 15th of October 2005

# INTRODUCTION

The idea for the conference which's papers are presented in this book was born during a Post-Doc-stage of Alexa Dufraisse at the Institute of Prehistory and Archaeological Science (IPAS) of the Basel University. Her occupation with Neolithic firewood made clear that although in different European countries there is many research on the field of anthracology, the scientific exchange is limited, especially concerning the French-speaking community and the German resp. English speaking (resp. publishing) scientists. This is a pity because especially in France, in the last decades a lot of fundamental investigations, above all also concerning methodological questions were performed which found only limited attention in the international, mainly English publishing scene.

The symposium was mainly dedicated to the evidence of firewood economy in (pre-)historic periods. Around twenty scientists from seven countries were assembled in Basel the 14-15 October 2004 to discuss the main scientific themes around firewood. These were manifold and covered a broad range of scientific research in anthracology. This is also mirrored in the papers in this book.

In the paper of C. Delhon the possibilities and limitations of charcoal investigations from pedological profiles are presented (off-site data). She shows that - if the appropriate methodological basics are respected - it is possible to reconstruct very well the development of the vegetation during the last several thousand years with such charcoal.

The paper of I. Stuijts presents different strategies for the sampling of charcoal and its subsequent preparation from recently excavated sites in Ireland.

The first paper of V. Py concerns the firewood economy in medieval silver mines in the SW-Alps. She could show that large amounts of wood were used to set fire for breaking down the hard quartzite bedrock. The results allow the reconstruction of the fuel management and the use of the environment. One of the aims of the study is methodological approaches (see below, the second paper by V. Py).

Of particular interest is the two following contributions. In the first one, A. Dufraisse presents a review of the potential of the wood anatomical features to improve, ecologically and economically, the interpretation of charcoal spectra.

Then, she develops a method for the reconstruction of wood diameter, based on a theoretical model. She also presents a new approach which combine the tree-ring bending and the radial growth.

In the same way, the contribution of T. Ludemann presents a standardized anthracological method for the reconstruction of the wood-diameters analysed. Basis for this are the remains of modern charcoal-burning. The method can be applied for historic samples and is a sensitive tool to establish qualitative and quantitative information about the wood taxa compositions and diameters used in past charcoal production.

Results of an experimental fire-setting are also presented in the second paper of V. Py, together with B. Ancel. They try to rediscover the technical know-how and evaluate the combined role the fires intrinsic (fuel) and exterior (ventilation, pyre architecture) factors.

The paper of Y. Carrión Marco presents the results of the dendrological analysis of carbonised material from a final Neolithic (Bell Beaker) dolmen, and shows that predominantly one wood species was used for the construction of the wooden parts of the monument.

The two last papers, written by S. Thiébault and V. Bernard (with the collaboration of S. Renaudin and D. Marguerie) deal with possibilities for evidencing branch cutting (pollarding, trimming) in the tree-ring-sequences of firewood. Such studies form a very important base for e.g. evidencing animal foddering, but also branch-cutting for firewood.

All in all, this book gives an excellent overview about the ongoing research on the field of anthracology, mainly firewood research, in Europe. It is hoped that it enhances the scientific exchange between the French (incl. Other "Roman" languages), German and English speaking communities.

Stefanie Jacomet

IPAS, University of Basel

# PALAEO-ECOLOGICAL RELIABILITY OF
# PEDO-ANTHACOLOGICAL ASSEMBLAGES

Claire DELHON

MAE, Protohistoire européenne, Archéobotanique et paléo-écologie
UMR 7041 ArScAn CNRS, Université de Nanterre Paris X
21 allée de l'Université F-92023 Nanterre (France)
claire.delhon@mae.u-paris10.fr

**ABSTRACT:** Fragments of wood charcoal produced by vegetation fire are often found in natural pedosedimentary sequences and are potentially valuable palaeoenvironmental proxies. However, palaeo-ecological reliability of pedo-anthracological assemblages, which often contain a small number of fragments, is sometimes questioned. The important pool of data obtained through rescue excavations (more than 40000 fragments) has allowed for the first time a statistical approach of this issue. It appears that the biodiversity of pedo-anthracocenoses is always great in relation to the small number of fragments. Even if the ideal sample size is over 250 fragments, we demonstrate that for far smaller samples the species diversity of assemblages mainly depends on the original vegetation, and is very little influenced by "deposition", taphonomy", "sampling" and "identification" factors. The phytosociological hierarchy of the taxa is well respected (leader species, minor species, rare species...). In particular, the behaviour of rare species, whose apparition is not linked to the number of fragments in the sample, shows that pedo-anthracological assemblages record well vegetation changes.
KEY WORDS: pedo-anthracology, palaeoenvironment, off site sampling, sample size, palaeoecological reliability

**RÉSUMÉ:** Les charbons de bois issus d'incendie de la végétation sont fréquemment retrouvés dans des séquences pédosédimentaires naturelles, et présentent un fort potentiel pour l'étude des paléoenvironnements. Toutefois, la validité paléo-écologique des assemblages pédo-anthracologiques, souvent constitués d'un nombre réduit de fragments, est parfois mise en doute. Grâce à un important corpus de données issues de l'archéologie préventive (plus de 40000 fragments), une approche statistique de ce problème est mise en œuvre pour la première fois. Il apparaît alors que la biodiversité des pédo-anthracocénoses est toujours grande, par rapport au faible effectif analysé. Bien que l'effectif idéal se situe au delà de 250 fragments, on montre que, même avec des effectifs beaucoup plus faibles, le facteur qui influe le plus sur la diversité taxonomique des assemblages pédo-anthracologiques est la diversité taxonomique de la végétation source, les facteurs "dépôt", "taphonomie", "prélèvement" et "identification" ayant une importance moindre. Les "rangs" phytosociologiques des taxa sont respectés (espèces chefs de file, secondaires, rares...). En particulier, le comportement des espèces rares, dont l'apparition ne semble pas liée à l'effectif, démontre que les assemblages pédo-anthracologiques enregistrent de façon satisfaisante les changements de végétation.
MOTS-CLÉS: pédo-anthracologie, paléoenvironnement, prélèvement hors site, effectif de l'échantillon, représentativité paléoécologique

**ZUSAMMENFASSUNG:** Holzkohlen von Vegetationsbränden werden oft in natürlichen Bodenprofilsequenzen gefunden und bieten ein hohes Potential für das Studium der früheren Umwelt.Dennoch wird an der paläoökologischen Gültigkeit der oft nur durch eine geringe Anzahl repräsentierten Holzkohlefunde aus Bodenprofilen manchmal gezweifelt.Eine wichtige Datenmenge (von mehr als 40000 Fragmenten) wurde durch Rettungsgrabungen erhalten und erlaubt erstmals eine statistische Annäherung an diese Problematik. Es scheint, dass die Biodiversität der Holzkohlereste in Bodenprofilen gegenüber der geringen Anzahl der Fragmente immer gross ist. Auch wenn die ideale Probengrösse mehr als 250 Fragmente umfasst, zeigen wir, dass bei weit kleineren Proben die Artenvielfalt hauptsächlich von der Ausgangsvegetation abhängt und nur geringfügig durch die Faktoren "Ablagerung", "Taphonomie", "Probenentnahme" und "Bestimmung" beeinflusst wird. Die pflanzensoziologische Hierarchie der Taxa wird eingehalten (Zeigerarten, sekundäre Arten, seltene Arten...). Das Verhalten seltener Arten, deren Vorkommen nicht an die Anzahl der Fragmente gebunden ist, zeigt, dass Holzkohlefunde aus Bodenprofilen Vegetationsveränderungen gut widerspiegeln *(Translation Danièle Martinoli and Britta Pollman)*.
STICHWORTE: Pedo-anthracologie, frühere Umwelt, Probenentnahme off-site, Probengrösse, paläoökologische Repräsentativität

# INTRODUCTION

If methodological studies have established the legitimacy of palaeo-ecological interpretations based on archaeo-anthracological studies (see particularly CHABAL 1992, 1997), it is not the same for pedo-anthracology. For

the last 15 years, a group of researchers stemming from the *Institut Méditerranéen d'Ecologie et de Paléoécologie* of Marseille, France (M. Thinon, C. Carcaillet, B. Talon notably), has tried to lay down the methodological foundations for the discipline, without completely succeeding in removing the suspicion it

1: Châteauneuf/Isère, Beaume
2: Montélier, Claveysonne
3: Chabeuil, les Brocards
4: Montvendre, les Châtaigniers
5: Upie, les Vignarets
6: Crest, Bourbousson
7: Chabrillan, Saint-Martin
8: La Roche/Grane, les Trélayes
9: Roynac, les Roches
10: Roynac, le Serre
11: Bonlieu/Roubion, les Bardes
12: Montboucher, les Hayes
13: Montboucher/Jabron, Constantin
14: Espeluche, Lalo
15: La Garde-Adhémar, Surel
16: Pierrelatte, les Malalones
17: La Palud, les Girardes
18: La Palud-les Dèves/les Bouchardes
19: Bollène, les Bartras
20: La Palud, les Petites-Bâties
21: Bollène, Pont de la Pierre
22: La Motte-du-Rhône, le Chêne
23: Mondragon, les Juilléras / La Motte-du-Rhône, la Prade
24: Mondragon, le Duc
25: Mondragon, les Brassières
26: Mondragon, les Ribauds
27: Orange-Q600
28: Caderousse, les Crémades
29: Caderousse, les Négades
30: Caderousse, Saint-Pierre

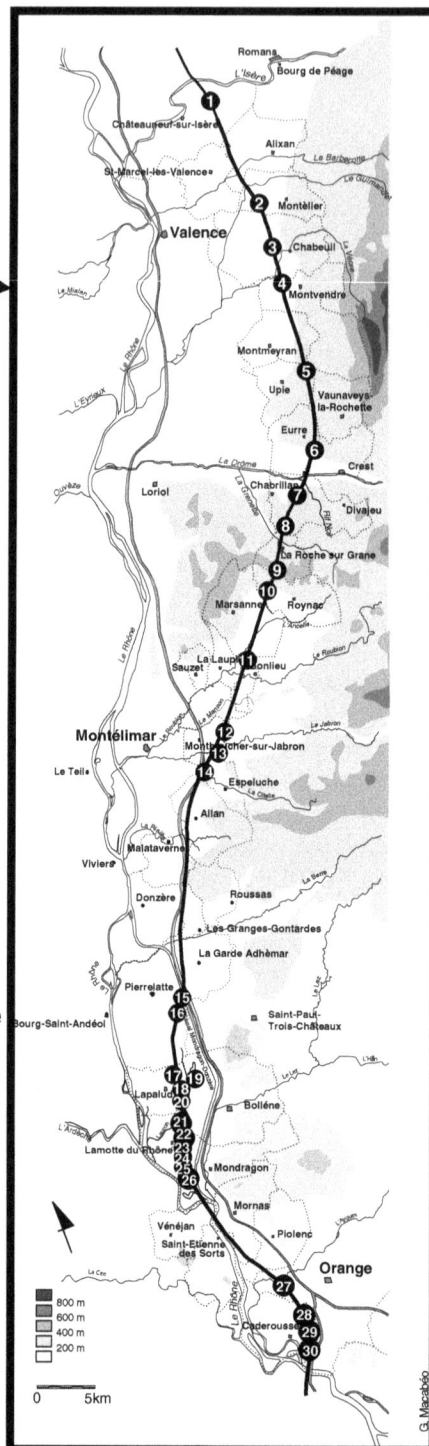

**Figure 1.** Main sampling-sites of the TGV-Méditerranée excavations used for this study.

inspires to a part of the scientific community. Critics mainly concern the size of pedo-anthracological assemblages, their palaeo-ecological reliability, and the spatial and temporal scales involved in the analyses. In this paper, we deal with the first two problems, the third still needs to be investigated further. The great number of pedo- and archaeo-anthracological samples supplied by the TGV-Méditerranée excavations provides a valuable pool of data to reply to these criticisms. The comparisons between 744 charcoal assemblages from actual archaeological sites and 390 charcoal assemblages from natural sequences located in the same area, enable us to prove the palaeo-ecological reliability of the reconstructed plant spectra, in spite of the low number of fragments in each assemblage.

## DEFINITION

"Pedo-anthracology" is a twenty years old discipline (Thinon 1978, 1992, Carcaillet and Thinon 1996, Thinon and Talon 1998, Tardy 1998) that consists of the taxonomical determination of wood charcoal fragments from soils or buried soils, with the aim of reconstructing palaeoenvironments. By extension, the term is also used for the study of charcoal fragments recovered "off archaeological site", in colluvio-alluvial layers. Although the term of "geo-anthracology" should therefore be more accurate (Vernet 2002), "pedo-anthracology" is more commonly used.

## MATERIAL AND METHODS

The present study is based on the analysis of about 40 000 carbonised wood fragments[1] from archaeological structures and pedo-sedimentary sequences brought to light by excavations prior to the construction of the TGV-Méditerranée fast-railway (Fig. 1). The aim of the charcoal analysis was the reconstruction of the vegetal landscape from the Late Glacial (Delhon 2005, Delhon et al. forthcoming), but it provided enough data to allow a more methodological study of the palaeo-ecological reliability of "off-site" charcoal assemblages.

## Sampling

Ten kilograms of sediment have been sampled from every stratigraphic layer, whether it presented or not macroscopical charcoal, and sifted through a 500 µm mesh. Systematic sampling provided charcoal from layers thought to be sterile. Some layers were actual fire layers, containing macroscopical charcoal and bearing features of heating (soil rubification), but carbonised wood fragments have been also found in layers without any evidence of *in situ* vegetation fire.

## Identification

Despite the small size of the fragments from natural sequences (500 µm-1 mm), the taxonomical determination is done in the same way as for "archaeo-charcoal". The fragments are manually broken, ideally according to the three characteristic anatomical planes (Western 1963), and observed under optical reflection microscope. Observed anatomical features are compared to those described in wood anatomy atlases (Schweingruber 1990, Vernet et al. 2001) and to reference material (reference collection of the *Laboratoire d'Archéobotanique et de Paléoécologie*, UMR 7041, Nanterre).

In practice, fragments are often too small to be broken along these three planes. A first observation at a small magnification can help to orientate them according to the most favourable plane to break and observe. At the same time, some anatomical criteria can be noted (vessels arrangement, ray width...) in order to avoid the loss of information due to breaking. Some characteristic features can only be observed on a relatively wide surface (in particular the arrangement of the various anatomical components on the transversal plane), some others need repeated observations (ray width and type, mean number of bars in scalariform plates...). This kinds of information is scarcely legible on very small fragments, in spite of their being at the root of the usual determination keys. On the contrary, the observation of features usually regarded as "additional" or even sometimes "incidental" information can be decisive (even if not self-sufficient) in the identification of a species. For example, *Salix* sp. is often recognized thanks to its large punctuations, ash (*Fraxinus* sp.) thanks to its pores arrangement, included the smallest in final wood, in "8-shaped" clusters, and Pomoideae thanks to their regular round-shaped ray-cells seen on tangential section (Fig. 2). In spite of the small size of the fragments, the determination almost always reaches at least the family level, very often the genus level, and sometimes the species level.

## PALAEO-ECOLOGICAL RELIABILITY OF PEDO-ANTHRACOLOGICAL ASSEMBLAGES

During a fire, part of the vegetation is carbonised (Fig. 3). Charcoal gets recovered and undergoes several taphonomical processes. The pedo-anthracological sample that is studied only represents a fraction of the total anthracocenosis. Thus, the reliability of palaeoenvironemental interpretations made from that kind of material depends on to what extend:
- the charcoal sample is representative of the total anthracocenosis;
- the anthracocenosis is representative of the burnt vegetation;

- and the burnt vegetation is representative of the whole vegetation cover.

As they are often made of a small number of fragments, pedo-anthracological assemblages can only contain a limited number of species. Moreover, and contrary to archaeological charcoal, the studied fragments have not been collected from a more or less wide area ("site catchments" or "supply area" theories: VITA FINZI AND HIGGS 1970, GEDDES 1987, CHABAL 1997, DUFRAISSE 2002) the assemblage formed "naturally" during a vegetation fire. In such conditions, we do not know if the charcoal is well mixed, or if the pedo-anthracological assemblage has a high probability of containing only wood from a sole tree, that burnt in the place were the sample is taken. Some anthracologists, referring to a sifting mesh of 4 mm and to archaeological charcoal, usually advocate that samples should contain 400 to 500 fragments, and they consider that assemblages made of less than 250 to 300 fragments are far less valuable (CHABAL 1997, CHABAL ET AL. 1999). Following these requirements, pedo-anthracological assemblages would only exceptionally be reliable, and the volume of sediment that should be sifted would discourage even the most motivated anthracologists!

## Biodiversity of anthracocenoses

The number of identified taxa in an assemblage is always lower than (or scarcely equal to) the number of analysed fragments. Thus, the low mean number of fragments in the pedo-anthracological assemblages restricts their floristic diversity. In any case, the ratio of the number of identified taxa versus the number of analysed fragments is higher than expected. At the beginning of this study, we thought that charcoal fragments sampled off archaeological sites would be less accurate tracers of the environment's biodiversity than archaeological charcoals. But the comparison of the two types of data shows an opposite result (FIG. 4 and 5).

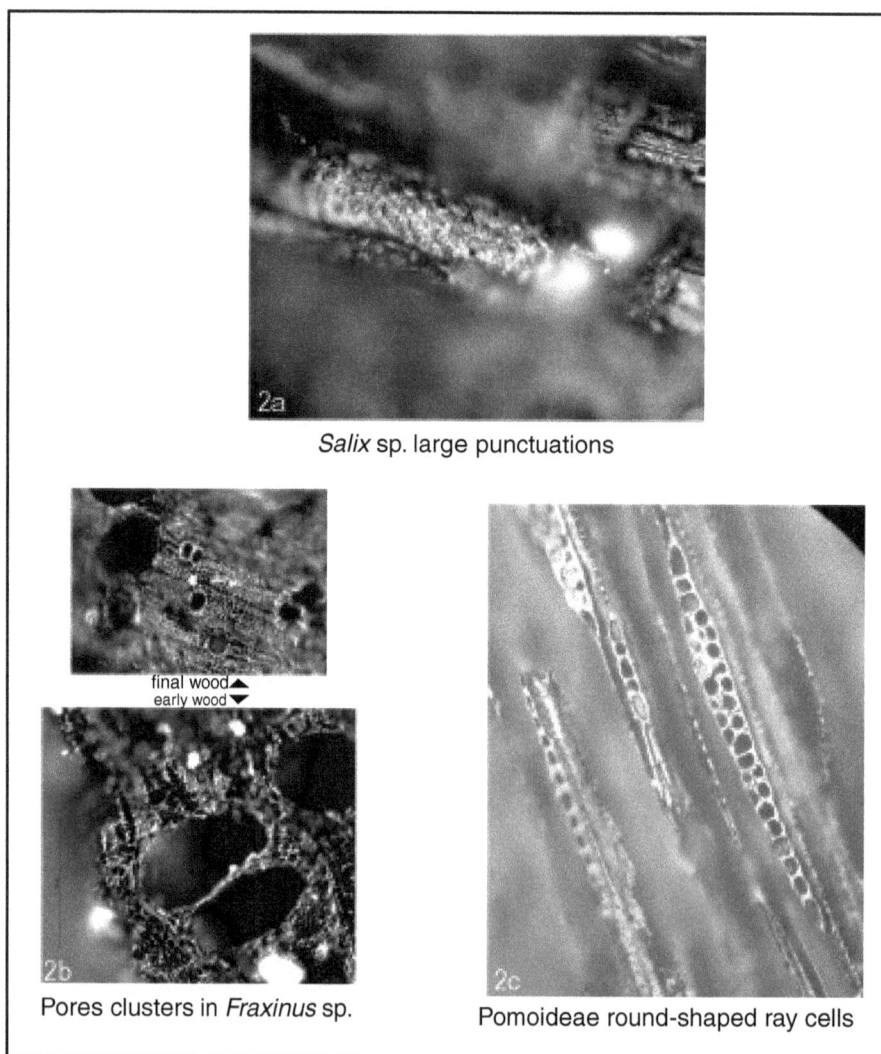

Salix sp. large punctuations

final wood▲
early wood▼

Pores clusters in Fraxinus sp.

Pomoideae round-shaped ray cells

**Figure 2.** Some anatomical criteria often used in pedo-anthracology (all photos show fossil charcoal).

In the context of rescue excavations, it is often difficult to implement the sampling protocols recommended by specialists, and as a result the number of identified taxa is always low, on archaeological sites as well as in natural sequences. However, the pedo-anthracological assemblages are more diversified (mean: 2.8 taxa per assemblage) than archaeological samples (mean: 2 taxa per assemblage, more than a half of the samples provided a single taxon). This is partly due to the occurrence, among archaeological assemblages, of many samples taken from charcoal concentrations in pits or hearths that often proved to be monospecific. But the low ecological representativity of archaeological charcoal is not the only reason for the difference between the two types of data. It appears clearly that pedo-anthracological samples are much richer than one would suppose according to the low number of fragments they are made of. The pedo-anthracological samples provide up to 18 different taxa, while archaeological samples do not contain more than 15 taxa.

It is possible that low concentration of charcoal in soils, that encourages cautious sifting, is indirectly responsible for the great diversity of pedo-anthracological samples,

while the larger size and the greater quantity of fragments in archaeological structures give a false impression of abundance that does not encourage more exhaustive sampling (for example extended to the entire occupation level). Moreover, large-scale sampling was often impossible, as mostly hollow structures were preserved.

However, the differences between the two types of data are not only due to a mediocre reliability of archaeological assemblages; it is obvious that the representativity of the spectra obtained with pedo-anthracological assemblages is high, in spite of their being small.

## How to quantify the proportion of each taxon?

The aim of pedo-anthracology is not only to propose a list of species but also to reconstruct the variation of their proportions in the vegetation, that can then be interpreted in terms of biogeography, climate changes and anthropogenic impact. In order to do this, it is necessary to quantify the absolute and relative frequence of each taxon.

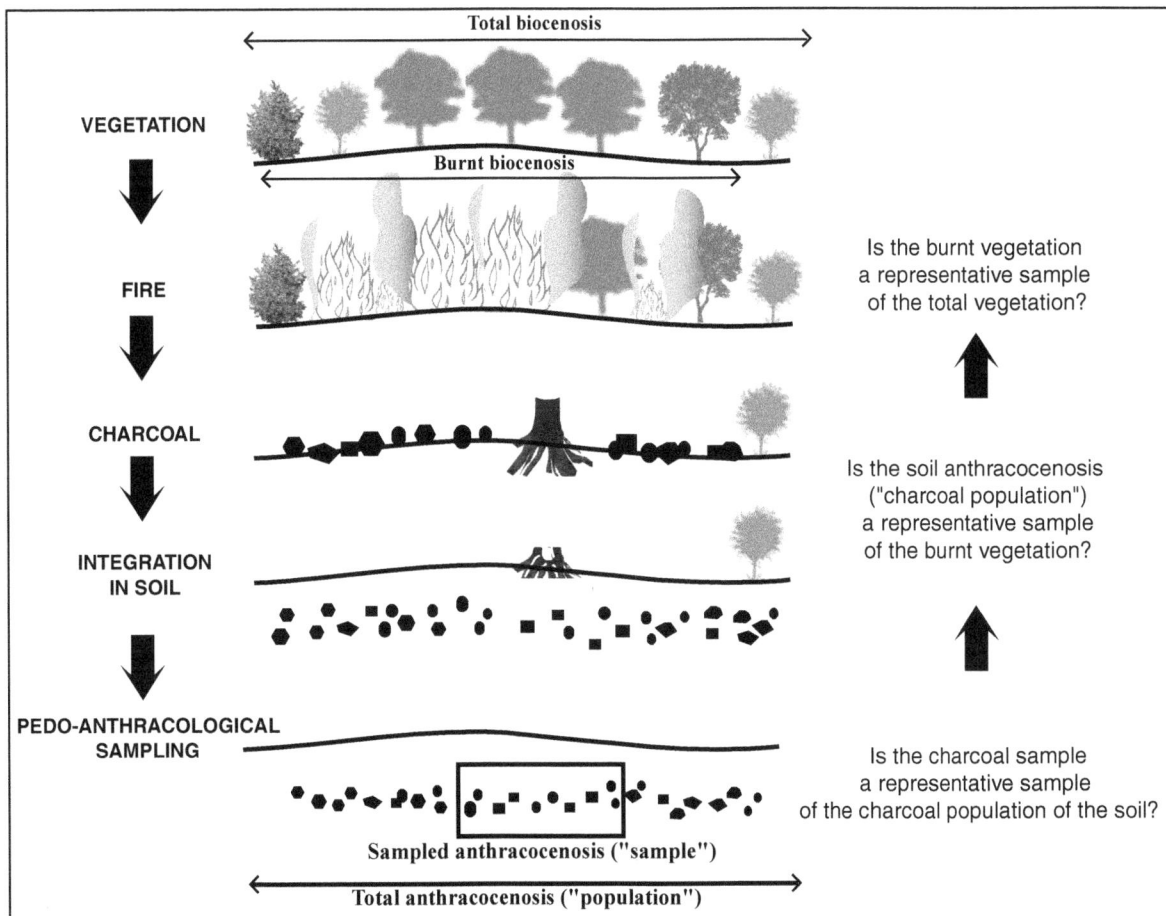

**Figure 3.** Process leading to the transformation of a living biocenosis into a fossil anthracocenosis, and related problems for the representativeness of the pedo-anthracological samples.

| | | Off site samples | | | | | | | | Archaeological samples | | | | | | | |
|---|---|---|---|---|---|---|---|---|---|---|---|---|---|---|---|---|---|
| | | Epipal. | Meso. | Néolithic | Metal ages | Classical / Middle Ages | modern | uncertain | TOTAL | Epipal. | Meso. | Néolithic | Metal ages | Classical / Middle Ages | modern | uncertain | TOTAL |
| | 1 | 0,8 | 3,2 | 5,3 | 5,8 | 17,4 | 4,2 | 0,8 | **37,5** | 0,1 | 0,5 | 16,3 | 31 | 5,2 | 0 | 0,3 | **53,4** |
| | 2 | 0,5 | 1,3 | 5 | 7,9 | 8,2 | 2,4 | 0 | **25,3** | 0,3 | 0,4 | 8,7 | 12,5 | 2,3 | 0 | 0 | **24,2** |
| | 3 | 0,5 | 0 | 3,4 | 1,3 | 6,8 | 0,8 | 0 | **12,8** | 0,1 | 0,8 | 2,2 | 6,3 | 1 | 0 | 0 | **10,4** |
| | 4 | 0 | 0 | 1,3 | 1,8 | 5 | 1,3 | 0 | **9,4** | 0 | 0,5 | 1,7 | 1,5 | 1,4 | 0 | 0 | **5,1** |
| | 5 | 0 | 0 | 0,3 | 1,1 | 1,6 | 0,3 | 0 | **3,3** | 0 | 0 | 0,4 | 0,6 | 1,3 | 0 | 0 | **2,3** |
| number of identified taxons | 6 | 0 | 0 | 0,3 | 0,5 | 2,1 | 0,3 | 0 | **3,2** | 0 | 0 | 0,1 | 0,4 | 0,9 | 0 | 0 | **1,4** |
| | 7 | 0 | 0 | 0 | 0,3 | 1,8 | 0 | 0,3 | **2,4** | 0 | 0 | 0 | 0 | 0,5 | 0 | 0 | **0,5** |
| | 8 | 0 | 0 | 0 | 0,3 | 1,3 | 0 | 0 | **1,6** | 0 | 0 | 0 | 0 | 0,6 | 0 | 0 | **0,6** |
| | 9 | 0 | 0 | 0,3 | 0,3 | 0,5 | 0 | 0 | **1,1** | 0 | 0 | 0 | 0 | 0,4 | 0 | 0 | **0,4** |
| | 10 | 0 | 0 | 0 | 0 | 0,5 | 0 | 0 | **0,5** | 0 | 0 | 0 | 0 | 0,1 | 0 | 0 | **0,1** |
| | 11 | 0 | 0 | 0 | 0,3 | 0,5 | 0 | 0 | **0,8** | 0 | 0 | 0 | 0 | 0,5 | 0 | 0 | **0,5** |
| | 12 | 0 | 0 | 0 | 0,3 | 0,3 | 0 | 0 | **0,6** | 0 | 0 | 0 | 0 | 0,4 | 0 | 0 | **0,4** |
| | 13 | 0 | 0 | 0 | 0 | 0,8 | 0,3 | 0 | **1,1** | 0 | 0 | 0 | 0 | 0,1 | 0 | 0 | **0,1** |
| | 14 | 0 | 0 | 0 | 0 | 0,3 | 0 | 0 | **0,3** | 0 | 0 | 0 | 0 | 0,3 | 0 | 0 | **0,3** |
| | 15 | 0 | 0 | 0 | 0,3 | 0 | 0 | 0 | **0,3** | 0 | 0 | 0 | 0 | 0,1 | 0 | 0 | **0,1** |
| | 16 | 0 | 0 | 0 | 0 | 0 | 0 | 0 | **0** | 0 | 0 | 0 | 0 | 0 | 0 | 0 | **0** |
| | 17 | 0 | 0 | 0 | 0 | 0 | 0 | 0 | **0** | 0 | 0 | 0 | 0 | 0 | 0 | 0 | **0** |
| | 18 | 0 | 0 | 0 | 0,3 | 0 | 0 | 0 | **0,3** | 0 | 0 | 0 | 0 | 0 | 0 | 0 | **0** |
| **TOTAL** | | 1,8 | 4,5 | 15,9 | 20,5 | 47,1 | 9,6 | 1,1 | **100%** | 0,5 | 2,2 | 29,4 | 52,3 | 15,1 | 0 | 0,3 | **100%** |

**Figure 4.** Frequency of the assemblages as a function of the number of identified taxons.

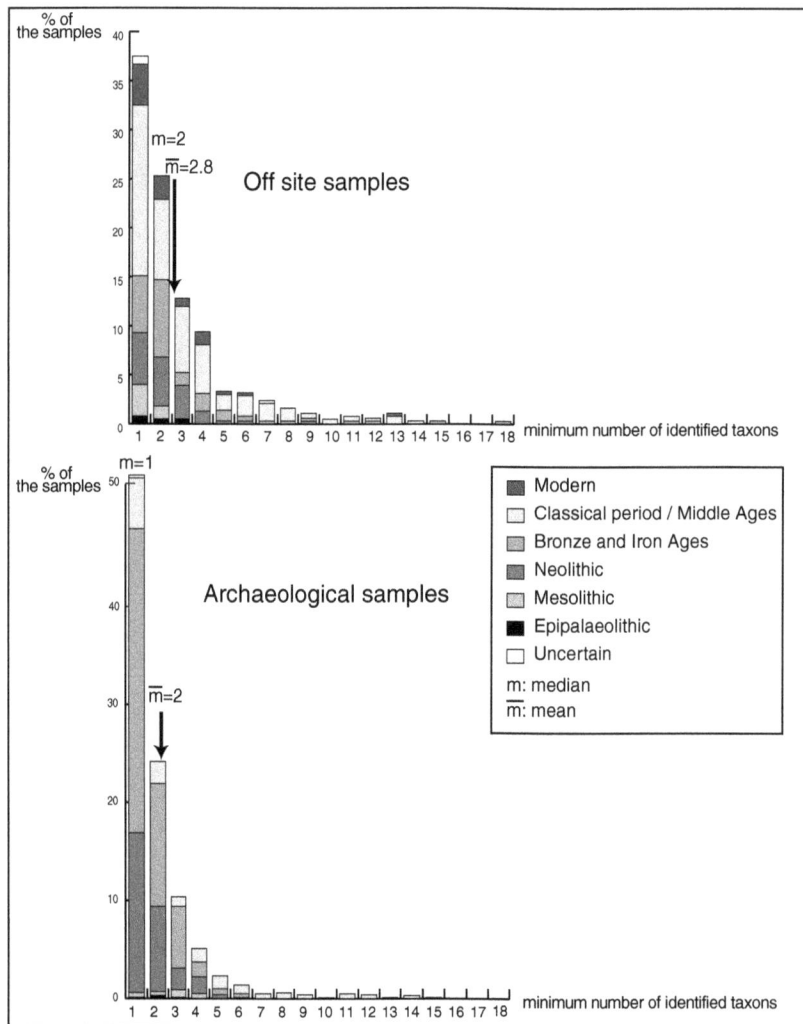

**Figure 5.** Frequency histograms of the samples in relation to the number of identified taxa. For off site samples, the mean number of taxa per assemblage is 2.8 and the median 2 (more than a half of the samples have provided 1 or 2 taxa only); for archaeological samples, the mean number of taxa per assemblage is 2, and the median is 1 (more than a half of the samples are monospecific). Note that some chronological periods are only represented by off site assemblages.

Weighing (KRAUSS MARGUET 1981) is a good way to take into account size differences between the fragments (as one can reasonably suppose that bigger fragments are the sign of a more important biomass), and counting is a good manner to take into account the quantity of fragments (as one can also suppose that more fragments are the sign of a more important biomass). This dilemma was solved about fifteen years ago for archaeological charcoal analysis, notably thanks to L. Chabal's studies (CHABAL 1990, 1992, 1997). When a big piece of charcoal breaks up it produces several small fragments, thus counting and weighing should not give different results. However, many anthracologists prefer counting, because there is no need for any particular equipement and it is faster and safer than weighing (the accuracy of scales can influence the results, especially in the case of very small fragments).

Even though this problem has been resolved for many years for archaeological charcoal, the question is still in suspense for pedo-anthracology. In France, researchers originating from the laboratory of Montpellier, historically more concerned by archaeo-anthracology, normally count the fragments, while those from the laboratory of Marseille, where pedo-anthraology was first developed, usually weigh the fragments. In that case, the variable used is "specific anthracomass" (THINON 1992). "Specific anthracomass", expressed in ppm (parts per million), is the ratio of the total mass of charcoal fragments extracted from a level (mg) versus the total mass of soil particles smaller than 5 mm (kg) (CARCAILLET AND THINON 1996). The sum of the mass of every charcoal belonging to a same taxon is the "specific taxonomical anthracomass" (AST). According to its defenders, the interest of this method is to allow "a quantitative and qualitative comparison of the taxa from different levels of a same soil, or different soils" (CARCAILLET AND TALON 1996: 235). Such a comparison supposes that the AST of various species are quantitatively similar for similar vegetations, and quantitatively different for different vegetations. But, we know that if mass reduction is equal for every species in given conditions of combustion, it varies with all the parameters other than the nature of fuel (intensity and duration of fire, humidity, amount of oxygen, etc.) (JUNEJA 1975, ROSSEN AND OLSON 1985 - synthesised in CHABAL 1997-, THÉRY-PARISOT 2001). Thus, it appears that AST is strictly comparable from one level to another and from one soil to another only if fire conditions have been strictly the same (season, wind, hygrometry, kind of fire…), which is almost impossible and at least indemonstrable. Moreover, every taphonomical process (splitting up, integration into the soil matrix by pedoturbation, biotic degradation, eluviation…) is able to make the AST vary, *a priori* in the same way for all the species, but in a different way from one place to another. Finally, the second term of the AST ratio, namely the mass of soil, can also vary for the same

volume according to the composition of the sediment, which can also be a problem for the comparisons.

Thus, we have chosen to count the fragments, considering that weighing is technically more difficult and doesn't offer any methodological advantages.

## Reliability of the relative representation of each taxon in pedo-anthracological assemblages as an indicator of the proportion in the total anthracocenosis

### *Mean "species-area" curve: about the sample size*

"Species-area" (or "effort-yield") curves allow a graphical estimation of the minimal size (number of fragments) of a representative sample (CHABAL 1982). For a given sample, it provides the minimal number of fragments that have to be analysed before one can consider that nearly all the ecological information is acquired, qualitatively speaking. When that number is reached, the identification of a new species (if there is any) needs the analysis of a large amount of fragments, which is considered as not profitable enough.

In our study, the number of fragments to be analysed has not been a real issue, because all the fragments of every pedo-anthracological sample have always been analysed (due to the small sample size). However, in order to visualise the evolution of the number of recognised taxa in relation to the number of analysed fragments, we draw a "mean" species-area curve (FIG. 6 and 7). The mean number of identified taxa was calculated for each sample-size (1 fragment, 2 fragments, 3 fragments…), for the archaeological samples and for the pedo-anthracological ones.

Contrary to a "classical" species-area curve, the "order of appearance" notion is not respected. The number of identified taxa quoted in the figure is a mean and not an actual value (assemblages made of a single fragment have all provided a single taxon, but it is not always the same one). In any case, the graphical representation shows how the assemblages are sorted, from isolated fragments (monospecific samples) to "ideal" samples of n charcoal that would contain all the ligneous species of the vegetation, in order to understand how limiting is the size of samples for palaeo-ecological reconstructions. The first information is that, in the studied context, samples from natural sequences always provide a wider specific diversity than archaeological samples (for the same number of fragments). One can also note that the "ideal" sample-size is not reached for any of the two types of samples. There is no real plateau in the curve, even if the gradient is decreasing between 50 and 150 fragments. It is nevertheless important to notice that the category "> 200 fragments" is made of only few assemblages of various

| Archaeological samples | | | Off site samples | | |
|---|---|---|---|---|---|
| Number of fragments per assemblage | Mean number of taxon per assemblage | standard deviation | Number of fragments per assemblage | Mean number of taxon per assemblage | standard deviation |
| 1 | 1 | 0 | 1 | 1 | 0 |
| 2 | 1,2 | 0,4 | 2 | 1,4 | 0,5 |
| 3 | 1,4 | 0,6 | 3 | 1,6 | 0,6 |
| 4 | 1,5 | 0,7 | 4 | 1,9 | 1 |
| 5 | 1,2 | 0,4 | 5 | 2,3 | 1 |
| 6 | 1,7 | 0,8 | 6 | 2,1 | 1 |
| 7 | 1,8 | 1 | 7 | 2,1 | 0,7 |
| 8 | 1,6 | 0,8 | 8 | 2,3 | 1,4 |
| 9 | 1,8 | 0,8 | 9 | 2,3 | 0,5 |
| 10 | 1,5 | 0,7 | 10 | 2,2 | 1,4 |
| 11 | 1,6 | 0,5 | 11 to 13 | 2,4 | 1,1 |
| 12 | 1,8 | 0,8 | | | |
| 13 | 2,5 | 1,3 | | | |
| 14 to 16 | 2 | 1,1 | 14 to 19 | 2,8 | 1,3 |
| 17 to19 | 1,9 | 1,1 | | | |
| 20 to 29 | 2,2 | 1,1 | 20 to 29 | 3,3 | 2,2 |
| 30 to 49 | 2,2 | 1,2 | 30 to 9 | 4,1 | 2,5 |
| 50 to 99 | 2,7 | 1,5 | 50 to 99 | 5,1 | 2,6 |
| 100 to 199 | 2,6 | 2,4 | 100 to 199 | 6,7 | 4,8 |
| >200 | 7,7 | 3,9 | >200 | 8,3 | 4,2 |

**Figure 6.** Mean number of taxons per assemblage in relation to the number of fragments.

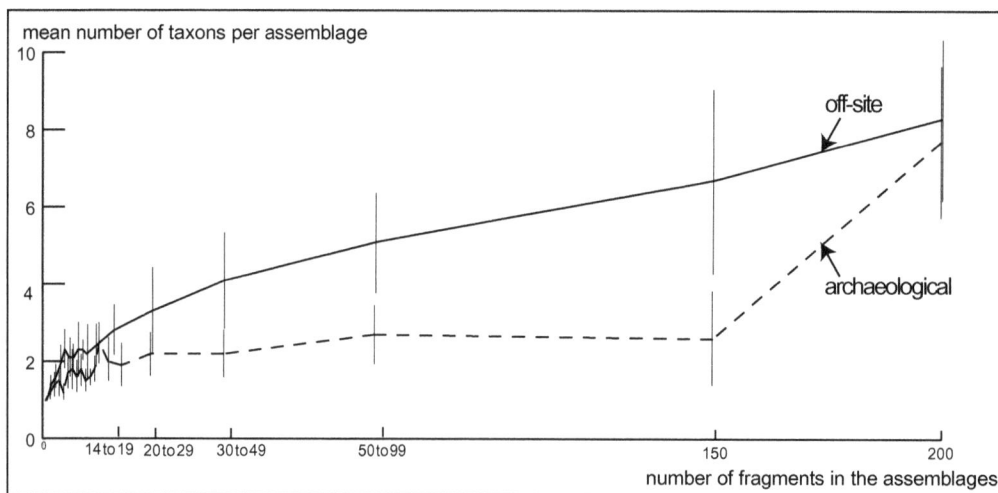

**Figure 7.** "Mean" species-area curves of off-site and archaeological (dots) anthracological assemblages from the TGV-Méditerranée excavations (vertical bars: standard deviations).

sizes (up to 926 fragments), among which some probably reach a sufficient size. Anyway, it appears clearly that 200 fragments is not, generally speaking, important enough to allow the identification of all the species that are present in the charcoal "population" of the soil (FIG. 3).

Pedo-anthracology also has the advantage of not being dependent on the existence of archaeological sites (FIG. 5). In connection with TGV-Méditerranée excavations, several periods provided very few (or even none) archaeological sites: Epipalaeolithic, Mesolithic, modern times, as well as shorter periods that are not individualised on Figure 4 (i.e. Middle Bronze Age). These periods have nevertheless provided sedimentary levels containing

charcoal, whose analysis supplied in the same time palaeo-ecological information about the vegetation and climate, and, in an indirect way, archaeological information on the relationship between societies whose cultural remains have not been exhumed yet, and their vegetal environment (agriculture, forest management, clearings…).

*Gini-Lorenz curves; proposal of a mode of calculation for relative frequencies*

The Gini-Lorenz curves can be used to describe the structure of plant associations. In temperate areas, it is usually observed that abondance-dominance phenomena lead to a stabilisation of species' frequencies with the

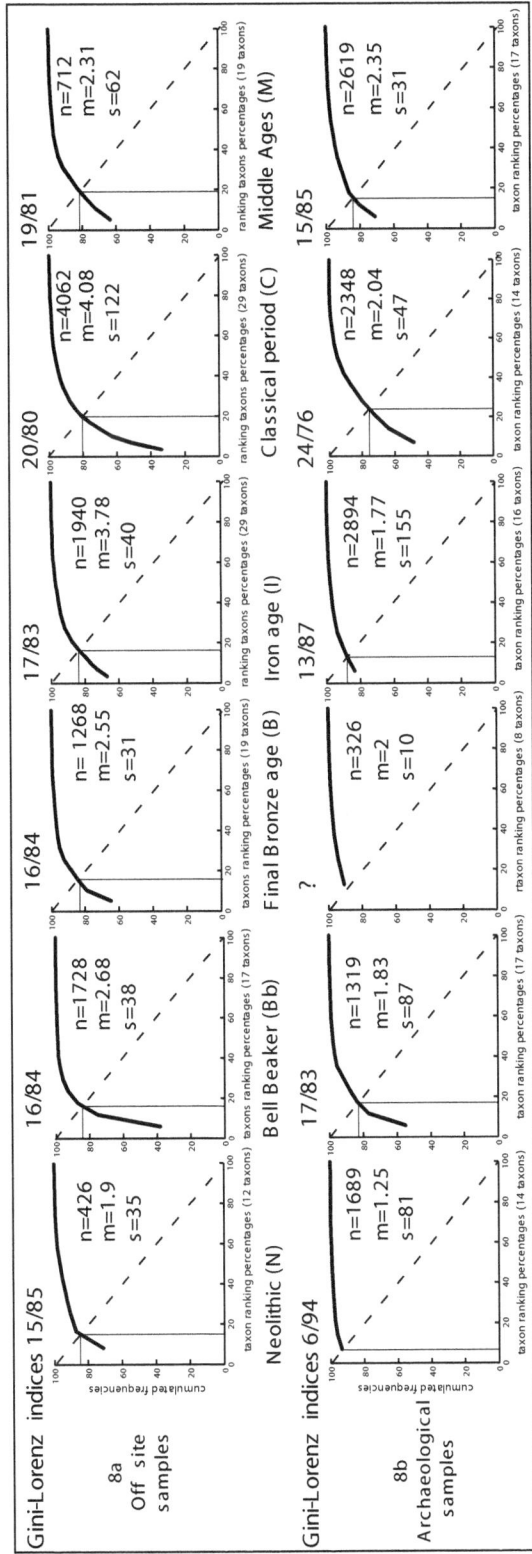

**Figure 8.** Graphical determination of Gini-Lorenz indices. N= total number of fragments (for all the period),s=number of samples, m=mean number of taxon per sample (for the considered period).

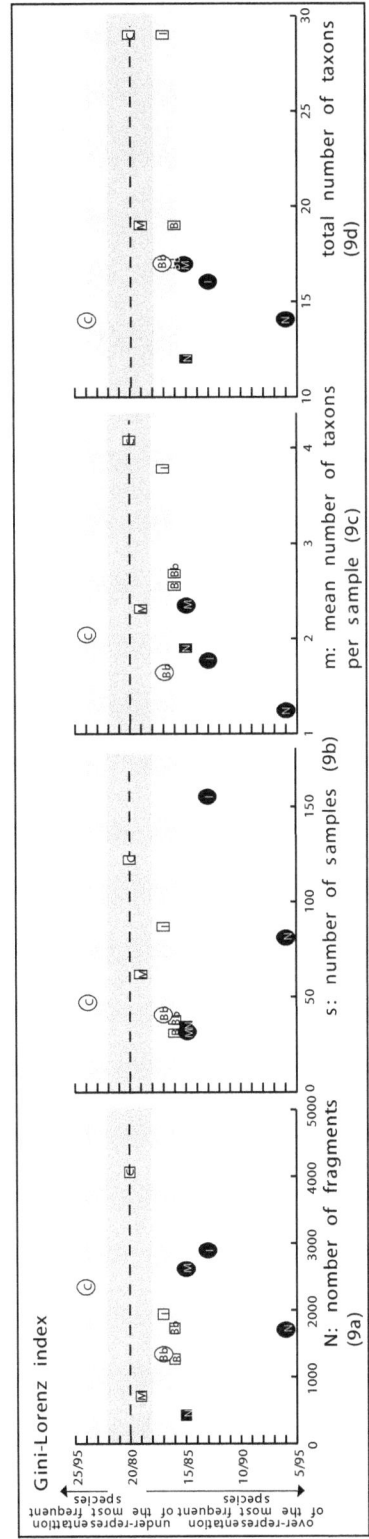

**Figure 9.** Variations of Gini-Lorenz indices in relation to the number of fragments (9a), the number of samples (9b), the mean number of taxons per sample (9c) and the total number of taxons (9d). White: indices close to 20/80, black: indices far from 20/80; square: off site, dot: archaeological samples.

20% most frequent species representing 80% of the total biomass, and the 80% less frequent species representing 20% of the biomass. In that case, the Gini-Lorenz index is quoted 20/80. This value is mostly observed in temperate areas when the vegetation is in a state of equilibrium. In case of perturbation, the index can decrease (i.e. 10/90), showing that the vegetation is temporally in a state of disequilibrium (CHABAL 1997). If the relative proportion of each taxon is identical in the anthracological assemblage and in the vegetation-source, the Gini-Lorenz index is the same. That proposition is the base of the demonstration of the good reliability of archaeo-anthracological assemblages as palaeo-vegetations tracers, quantitatively

speaking, in temperate areas (CHABAL 1997) as well as in tropical regions (SCHEEL-YBERT 2002).

Most of the samples involved in this study are too small to allow the drawing of a Gini-Lorenz curve. Nevertheless, it has been possible to trace these graphs for the sum of all the samples of the same period (FIG. 8). For off-site samples (FIG. 8a), the index is always good to very good. It varies from 15/85 to 20/80. If we assume that these anthracocenoses stem from vegetations at equilibrium, they appear to be globally representative of the abundance-dominance relationships inside these vegetations, in spite of the small sizes of samples. The indices are very good

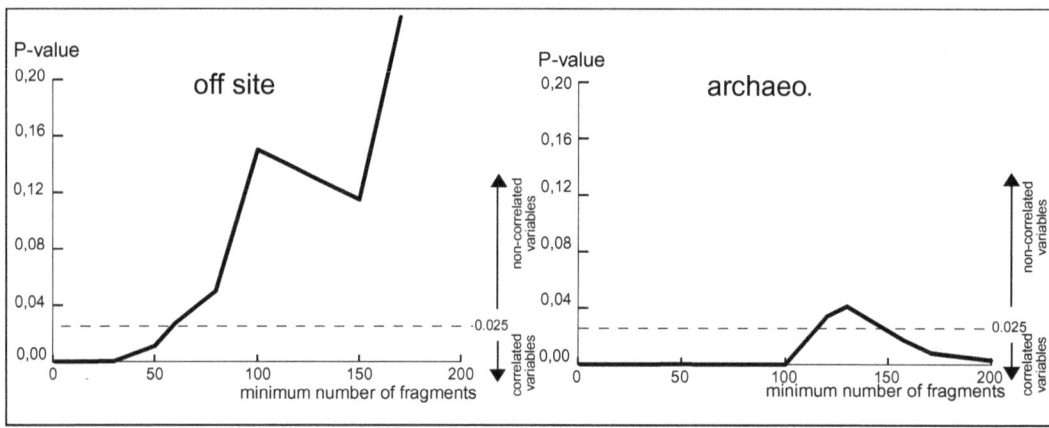

**Figure 10.** Evolution of Pearson's P-value in relation to the minimum number of fragments per sample. For off-site samples, the P-Value passes over 0.025 for samples of more than 60 fragments, which means that the number of identified taxons is independant of the number of identified fragments. On the contrary, for archaeological assemblages, the variables "number of fragments" and "number of taxons" stay correlated whatever the size of the sample is (that result would probably change if the samples passing over 200 fragments were more numerous).

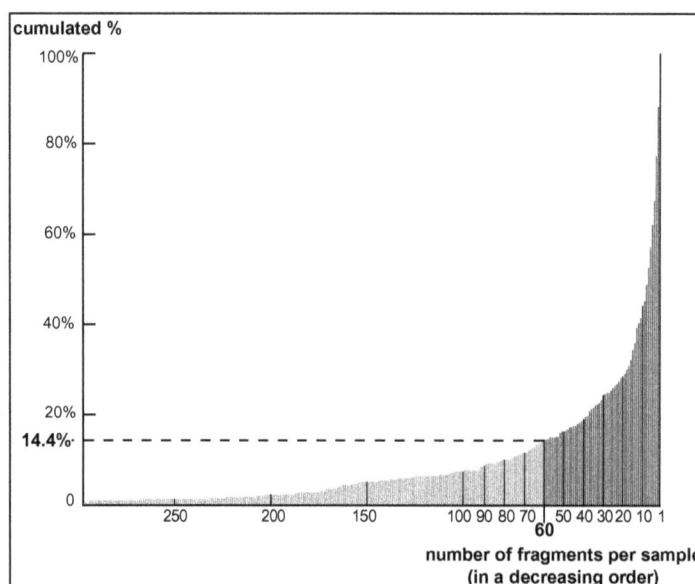

**Figure 11.** Cumulated frequencies of the TGV-Méditerranée samples in relation to their size. Samples of 60 fragments or more only represent 14.4% of the total.

for the Classical (20/80) and Medieval periods (19/81), and a bit weaker for the four other periods (15/85, 16/84 and 17/83), showing a slight over-representation of the more frequent species. Nothing indicates whether this over-representation is due to a sampling bias or to a disequilibrium in the vegetation itself.

The comparison of these indices with those obtained from archaeo-anthracological data shows better scores for pedo-anthracology, except for the Bell Beaker period (FIG. 8b), that is known to be a transition period for the vegetation in the area (DELHON 2005). Concerning the samples from archaeological structures, the index varies from 6/94 (very strong over-representation of the most-frequent species) to 24/76 (over-representation of the rarest species). For final Bronze Age, the low taxonomical diversity associated with a strong dominance of the most frequent taxon (deciduous oak, 90.8%) prevents the calculation of the index.

Grouping together all the samples from the same period is not sufficient to mitigate the low representativness of archaeological fragments. Graphs from Figure 9 show that indices do not get better when the number of analysed fragments is increasing, nor when the number of samples is increasing (FIG. 9a and 9b). On the other hand, the index is closer to 20/80 when there is an increase in the mean number of taxa per sample and/or in the total number of taxa (FIG. 9c and 9d). Figure 7 suggests that the number of taxa identified is linked to the number of analysed fragments, so these two parameters should influence the Gini-Lorenz index in the same way. In fact, it is obvious that the number of taxa is influenced by other parameters, and particularly by the actual number of taxa that take part in the vegetation (FIG. 2 and 9).

Pearson's coefficient of correlation between the number of fragments and the number of taxa in each sample was calculated for all the samples, for samples of more than 1 fragment, 10 fragments, 15, 30, etc. until more than 170 (off-site samples) or more than 200 (archaeological samples) fragments (FIG. 10). For archaeological samples, the number of identified taxa is always correlated to the number of analysed fragments, whatever the minimum sample size is (with only one exception for samples of more than 130 fragments, but it is not significant as the correlation exists when more or less samples are taken into account). This result shows that the bad values obtained for the Gini-Lorenz index are mainly due to the insufficient size of samples.

For pedo-anthracological samples, the correlation is less marked. Considering all the samples, the correlation does exist. On the contrary, if only samples of more than 60 charcoals are taken into account, the number of identified taxa is not correlated to the sample size anymore. This result means that for samples of more than 60 fragments the species diversity does not depend on the number of available fragments but on other factors, first of all the biodiversity of the source-vegetation itself.

These results confirm on the one hand the archaeo-anthracological prescriptions of an ideal minimal sample size of 250 to 300 fragments (CHABAL 1997, CHABAL ET AL. 1999). On the other hand, they show that for small samples off-site assemblages are more rapidly representative of the vegetation and that from samples of about 60 fragments the composition of the vegetation is the most important factor that influences the diversity of pedo-anthracological assemblages. In other words, pedo-anthracological samples seem to be more "homogenous" than archaeo-anthracological samples. That homogeneity is probably linked with taphonomical processes (in particular pedogenesis and accumulation) that lower the risk of mono-specific concentrations. Nevertheless, one can notice that pedo-anthracological samples of less than 60 fragments (and thus of insufficient size) are not rare in the studied corpus (only 14.4% of the samples pass over that number of fragments, FIG. 11).

Pedo-anthracological samples are always biased by an over-representation of the most-frequent species (namely deciduous oak, together with ash for the Bell Beaker period), which is translated into Gini-Lorenz indices a little bit under 20/80. These spectra result from the calculation of each species' percentage in relation to the total number of fragments (for each period). The great number of studied assemblages enables another mode of calculation of the relative values of the various species, based on the number of occurrences. The presence of a taxon in a sample has a value of 1, whatever the number of fragments of that taxon, and the absence of a taxon in a sample has a value of 0. The sum of presences/absences is done for each period, and is the base sum for the calculation of percentages (FIG. 12). Consequently, when a species represents 50% of the occurrences of a given period, it means that if a taxon is randomly chosen in a randomly chosen sample of that period, there is one chance in two to pick out this species.

Gini-Lorenz indices were calculated using the percentages of occurrences (FIG. 13). They systematically pass over those calculated from the percentages of fragments (FIG. 14). It means that the most-frequent taxa (namely deciduous oak) are under-representated (compared to their actual importance in the vegetation) to the benefit of other species. Anthracological diagrams based on the percentages of occurrences show a quantitative reliability of relative values, but they also give a better qualitative representation of the less-abundant taxa. The comparison with "classical" diagrams (percentages based on the number of fragments), together with the knowledge of the phytosociological behaviour of the species, make possible the correction of the distortion inherent to one or the other mode of calculation.

**Figure 12.** Principle of the calculation based on occurrences.

## Behaviour of the taxons in pedo-anthracological assemblages in relation to their phytosociological status

The diagram on Figure 15 represents the proportion of assemblages showing each taxon in relation to their total biodiversity. The percentage of samples containing each taxon was calculated in relation to the total number of taxa per sample.

The graph drawn for off-site samples (FIG. 15a) allows a classification of the taxa according to their behaviour, and the groups show a good phytosociological coherence:

- "leader-species" ("chefs de file", "tête de groupe", DA LAGE AND MÉTAILIÉ 2000) (deciduous oak, sclerophyllous oak, ash, elm, beech, willow/poplar, Scots pine) are those that characterise the plant associations. These species are dominant in undisturbed vegetations. Among those, deciduous oak is the "leader" of the main plant association of the middle Rhone valley (deciduous oak grove) and thus is nearly always present, whatever the size of samples (two thirds of all the samples, 70% of the monospecific samples, nearly all the samples of more than 10 taxa). The other "leaders" form nearly all the other monospecific samples, each one representing only a small percentage (less than 10%) of the samples. When the number of taxa increases, the frequency of "leaders" also increases, and they are finally represented in more than 80% (for the "leaders" of common plant associations: sclerophyllous oak, ash, beech) or in 50% to 70% (for the "leaders" of less frequent associations: elm, willow/poplar, Scots pine) of the assemblages that exceed 10 taxa.

- "ubiquitous secondary" taxa are Pomoideae and Prunoideae, both taxa gathering a great number of abundant but never dominant species ("taxon valise", CHABAL 1997), all heliophilous but adapted to various ecological conditions. Their curve on the diagram is similar to that of the "leaders".

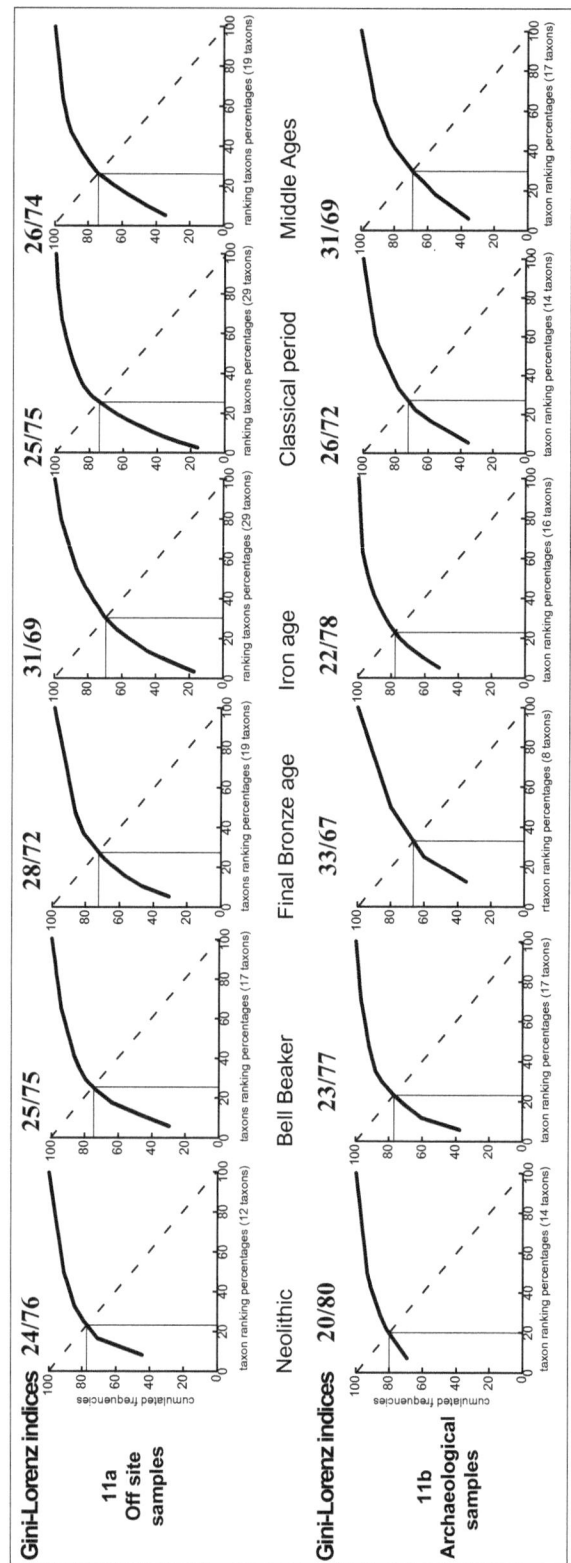

**Figure 13.** Graphical determination of Gini-Lorenz indices for frequencies calculation based on occurrences.

- "minor" taxa are often present in the plant associations, but they are neither dominant nor necessary to characterize these associations (maples, legumes, junipers, box, hazel,

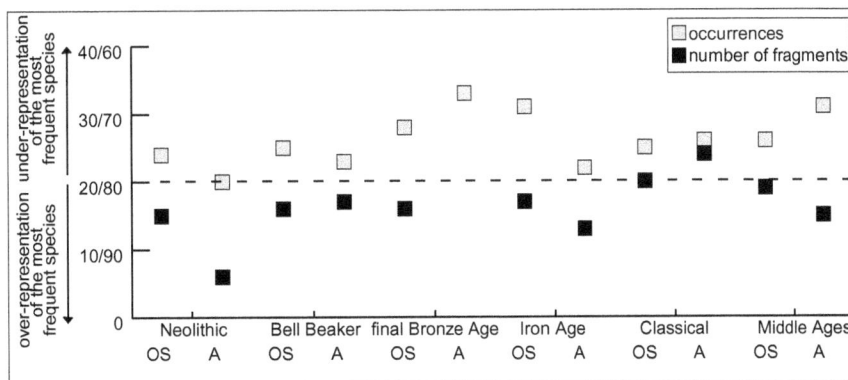

**Figure 14.** Gini-Lorenz indices with calculations based on the number of fragments (black) and on the number of occurrences (grey). OS= Off Site, A= Archaeological samples.

alder/birch, vine, *Rhamnus/Phillyrea*). They are most of the time absent from samples containing a small number of taxa (those are only made of "leader" species), but they rapidly appear (as soon as the assemblages reach 2 or 3 different taxa), and are present in an increasing proportion of samples (from 15 to more than 70%) when the sample's biodiversity increases.

- "intermittent" taxa (arbutus, pistachios, heathers, tamarisk, fir, holly, walnut) are the "secondary" taxa of the less frequent plant associations (mainly mesomediterranean and mesophile-humid associations). Some have been identified in small samples, but they generally occur in a significant proportion of the samples only when the diversity reaches 5 to 6 taxa. Their representation increases with the number of taxa, but they are nevertheless present in less then 40% of the assemblages of more than 10 taxa.

- "sparse" taxa (privet, elders, dogwood, olive, lime tree, *Cistus*) are not rare, but their density in plant associations is always low. They are identified nearly only in samples that offer a wide taxonomical spectrum. They correspond to the species that are the most dependent on combustion, taphonomical and sampling bias.

- "rare" taxa only occur occasionally and in very small quantities in plant associations (chestnut, clematis, ivy, ruscus, spindle-tree, alder buckthorn, mistletoe, Viburnum, Aleppo pine, yew, Chenopodiaceae). They are identified only in a very small proportion of the samples, but their occurrence seems not to be linked to the number of taxa in the sample. They appear "when they have to", whatever the biodiversity of the sample. This is an argument that the presence of a species in a pedo-anthracological sample does not depend primarily on the number of fragments or of taxa in the sample, but mainly on the presence of that species in the burnt vegetation. In other words, the "vegetation" factor works more than the "deposition", "taphonomy", "sampling" and "identification" factors to determine the composition of the pedo-anthracological

assemblage. Thus, pedo-anthracological samples are reliable proxies for reconstructing palaeo-vegetations.

The same diagram was drawn up for archaeological samples (Fig. 15b). The phytosociological status of the taxa is less well respected. This leads to a low representation (or even no representation at all) of the less frequent species ("intermittent", "sparse" and "rare"). This result confirms that obtained using Gini-Lorenz indices.

## CONCLUSIONS - INTERPRETATION OF PEDO-ANTHRACOLOGICAL SPECTRA: POSSIBILITIES AND LIMITS

Pedo-anthracological assemblages of at least 60 fragments are reliable indicators of past ligneous vegetation. Nevertheless, such assemblages are not the most frequent.

In case of smaller samples, the most frequent taxa in the assemblages are those that represent the most important biomass in the vegetation. The modern phytosociological and biogeographical data, the knowledge of each taxon's ecological requirements and tolerances as well as the consideration of previous data concerning the tendencies of the Holocene vegetal dynamics in the South of France (notably VERNET AND THIÉBAULT 1987, HEINZ AND THIÉBAULT 1998, TRIAT-LAVAL 1978, DE BEAULIEU 1977, ARGANT 1990, etc.) enable us to interpret even the smallest assemblages. The punctual observations made from a single small assemblage are often corroborated by the analysis of contemporary samples (DELHON 2005, DELHON *ET AL.* FORTHCOMING). Even if only dominant species are identified when the number of identified taxa is low, the punctual occurrence of rarer species does not usually go unnoticed, even if the whole biodiversity of the vegetation is not represented in the sample. Thus, it seems that every event that concerns the ligneous vegetation can be detected by pedo-anthracology, providing that a fire has allowed its fossilization.

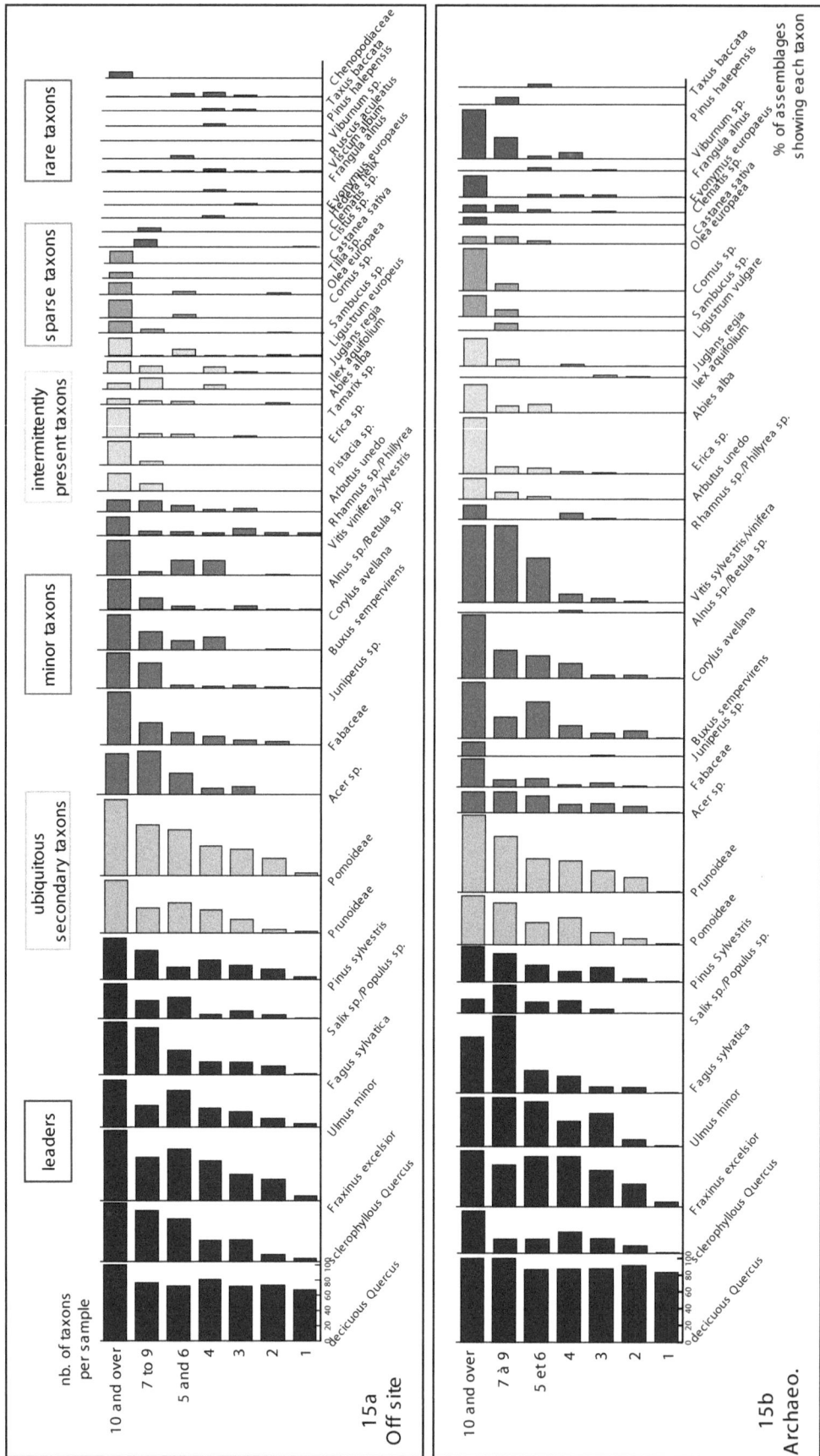

**Figure 15.** Proportion of samples showing each taxon in relation to the number of taxa per sample.

In any case, it is still impossible to reach an absolute quantitative approach of each specie's biomass. The use of AST for quantifying the taxa was rejected to the benefit of counting, but this method is not reliable for strict quantitative comparison from one soil to another. At best, when samples are big enough (> 60 fragments), it is possible to compare the proportions of the taxa, but it is still impossible to really know the density of the ligneous cover from pedo-charcoal only. We still have to use modern phytosociological and ecological data to estimate the density. That estimation is in any case only a subjective assessment and not a transfer function. The comparison with other data, more reliable for tracing non-ligneous vegetation (phytoliths, pollen, macroremains, pedological analyses…) is also helpful. This multi-proxy method has proved to be accurate in reconstructing the variations of the vegetation and of its management by past populations over the 15 last millennia in the middle Rhone valley (DELHON 2005, DELHON ET AL. FORTHCOMING).

---

[1] Charcoal analyses performed by the author and/or S. Thiébault, L. Fabre, L. Rousseau.

ACKNOWLEDGEMENTS
The anthracological data used in this study have been partly provided by S. Thiébault, L. Fabre and L. Rousseau, in the framework of the archaeological operation of the "TGV-Méditerranée". J.-F. Berger and J.-L. Brochier coordinated the environmental samplings. This work is also related to the ACI program *"Les paléoincendies: rythmes, origine et impact sur le fonctionnement des éco-anthroposystèmes dans le bassin méditerranéen nord-Occidental (Languedoc - vallée du Rhône - Côte d'Azur) depuis le Tardiglaciaire"* directed by J.-F. Berger, UMR 6031. The author thanks S. Thiébault for her useful collaboration and comments, and M. Tengberg and E. Hayes for their help in improving the English version.

REFERENCES

ARGANT J., 1990.- *Climat et environnement au Quaternaire dans le bassin du Rhone d'après les données palynologiques*, Lyon, Université Lyon 1, 199 p. (Documents du Laboratoire de Géologie de Lyon, 111).

DE BEAULIEU J.-L., 1977.- *Contribution pollenanalytique à l'histoire tardiglaciaire et holocène de la végétation des Alpes méridionales françaises*, Doctorat, Université Aix-Marseille III, 358 p.

CARCAILLET C., THINON M., 1996.- Pedoanthracological contribution to the evolution of the upper treeline in the Maurienne Valley (North French Alps): methodology and preliminary data, *Review of Palaeobotany and Palynology*, 91: 399-416.

CARCAILLET C., TALON B., 1996.- Aspects taphonomiques de la stratigraphie et de la datation de charbons de bois dans les sols: exemple de quelques sols des Alpes, *Géographie physique et Quaternaire*, 50 (2) : 233-244.

CHABAL L., 1982.- *Méthodes de prélèvement des bois carbonisés protohistoriques pour l'étude des relations homme-végétation (exemple d'un habitat de l'âge du Fer : le Marduel- St. Bonnet du Gard-fin VIIe-fin Ier siècle avant J.C.)*, DEA, Université de Montpellier II, 54 p.

CHABAL L., 1990.- L'étude paléo-écologique de sites protohistoriques à partir des charbons de bois: la question de l'unité de mesure. Dénombrement de fragments ou pesées ? *in:* Hackens T., Munaut A.V., and Till Cl. (eds), *Wood and Archaeology (Bois et Archéologie). First European Conference (Louvain-la-Neuve, October 2nd-3rd 1987)*, Strasbourg, PACT 22: 189-205.

CHABAL L., 1992.- La représentativité paléoécologique des charbons de bois archéologiques issus du bois de feu, *in:* J.-L. Vernet (ed.), *Les Charbons de bois, les anciens écosystèmes et le rôle de l'homme (Montpellier, décembre 1991)*, Paris, Editions Bulletin de la Société Botanique Française: 213-236 (Actualités Botaniques 1992-2/3/4).

CHABAL L., 1997.- *Forêts et sociétés en Languedoc (Néolithique final, Antiquité tardive). L'anthracologie, méthode et paléoécologie*, Paris, Editions Maison des Sciences de l'Homme, 189 p. (Document d'Archéologie Française, 63).

CHABAL L., FABRE L., TERRAL J.-F., THÉRY-PARISOT I., 1999.- L'anthracologie, *in:* C. Bourquin-Mignot *et al.* (eds), *La Botanique*, Paris, Editions Errance: 43-104 (Collection "Archéologiques").

DA LAGE A., MÉTAILIÉ G., 2000.- *Dictionnaire de Biogéographie végétale*, Paris, Editions C.N.R.S., 579 p. (C.N.R.S. dictionnaires).

DELHON C., 2005.- *Anthropisation et paléoclimats du Tardiglaciaire à l'Holocène en moyenne vallée du Rhône : études pluridisciplinairesdes spectres phytolithiques et pédo-anthracologiques de séquences naturelles et de sites archéologiques*, Doctorat, Université de Paris I, 843 p.

DELHON C., THIÉBAULT S., FABRE L., FORTHCOMING.- Evolution de la végétation d'après l'anthracologie, *in:* J.-F. Berger and J.-L. Brochier (eds), *Histoire des paysages et du climat de la fin des temps glaciaires à nos jours en moyenne vallée du Rhône d'après les données des travaux archéologiques du TGV-Méditerranée.*

DUFRAISSE A., 2002.- *Les habitats littoraux néolithiques des lacs de Chalain et Clairvaux (Jura, France): collecte du bois de feu, gestion de l'espace forestier et impact sur le couvert arboréen entre 3700 et 2500 av. J.-C. Analyses anthracologiques,* Doctorat, Université de Franche-Comté, 349 p.

GEDDES D., 1987.- Subsistance et habitat du Mésolithique final au Néolithique moyen dans le bassin de l'Aude (France), *in :* J. Guilaine (ed), *Premières communautés paysannes en méditerranée occidentale (Montpellier, avril 1983),* Paris, Editions C.N.R.S.: 201-207.

HEINZ C., THIÉBAULT S., 1998.- Characterization and palaeoecological significance of archaeological charcoal assemblages during late and post-glacial phases in southern France, *Quaternary Research,* 50: 56-68.

JUNEJA S.C., 1975.- Combustion of cellulosic material and its retardance - status and trends. Part 1: Ignition, combustion processes and synergism, *Wood Science,* 7: 201-208.

KRAUSS-MARGUET I., 1981.- Contribution à l'histoire de la végétation postglaciaire des grands causses d'après l'analyse anthracologique du gisement de la Poujade (Millau, Aveyron), *Paléobiologie continentale,* XII (1): 93-110.

ROSEN J., OLSON J., 1985.- The Controlled Carbonization and Archaeological Analysis of SE U.S. Wood Charcoals, *Journal of Field Archaeology,* 12: 445-456.

SCHEEL-YBERT R., 2002.- Evaluation of sample reliability in extant and fossil assemblages, *in:* S. Thiébault (ed.), *Charcoal analysis, methodological approaches palaeoecological results and wood-uses. Proceeding of the second international meeting of anthracology (Paris, september 2000),* Oxford, BAR Publishing: 9-16 (BAR International Series 1063).

SCHWEINGRUBER F.H., 1990.- *Anatomie europäischer Hölzer,* Bern und Stuttgart, Verlag Paul Haupt, 800 p.

TARDY C., 1998.- *Paléoincendies naturels, feux anthropiques et environnement forestiers de Guyane française du Tardiglaciaire à l'Holocène récent: Approches chronologique et anthracologique,* Doctorat, Université de Montpellier II, 343 p.

THÉRY-PARISOT I., 2001.- *Economie des combustibles au Paléolithique. Expérimentation, taphonomie, anthracologie,* Paris, Editions C.N.R.S. - C.E.P.A.M., 195 p. (Dossier de Documentation Archéologique, 20)

THINON M., 1978.- La pédoanthracologie: une nouvelle méthode d'analyse phytochronologique depuis le Néolithique, *Comptes Rendus de l'Académie des Sciences de Paris,* 287, série D: 1203-1206.

THINON M., 1992.- *L'analyse pédoanthracologique: aspects méthodologiques et applications,* Doctorat, Université d'Aix-Marseille III, 317 p.

THINON M. , TALON B., 1998.- Ampleur de l'anthropisation des étages supérieurs dans les alpes du sud: données pédoanthracologiques, *Ecologie,* 29 (1-2): 323-328.

TRIAT-LAVAL H., 1978.- *Contribution pollenanalytique à l'histoire Tardiglaciaire et Postglaciaire de la végétation de la basse vallée du Rhone,* Doctorat d'Etat, Université d'Aix-Marseille III, 343 p.

VERNET J.-L., 2002.- Preface, *in:* S. Thiébault (ed.), *Charcoal analysis, methodological approaches palaeoecological results and wood-uses. Proceeding of the second international meeting of anthracology (Paris, september 2000),* Oxford, BAR Publishing: V-VI (BAR International Series 1063).*

VERNET J.-L., THIÉBAULT S., 1987.- An approach to northwestern Mediterranean recent prehistoric vegetation and ecologic implications, *Journal of Biogeography,* 14: 117-127.

VERNET J.-L., OGEREAU P., FIGUEIRAL I., MACHADO YANES C., UZQUIANO P., 2001.- *Guide d'identification des charbons de bois préhistoriques et récents. Sud-Ouest de l'Europe: France, Péninsule Ibérique et îles Canaries,* Paris, Editions C.N.R.S., 395 p.

VITA FINZI C., HIGGS E.S., 1970.- Prehistoric economy in the Mount Carmel area of Palestine, site catchment analysis, *Proceedings of the Prehistoric Society,* 36: 1-37.

WESTERN A.C., 1963.- Wood and charcoal in archaeology, *in:* D. Brothwell, E. Higgs and G. Clark (eds.), *Science in archaeology: a comprehensive survey of Progress and Research,* New-York, Thames & Hudson: 150-158.

# CHARCOAL SAMPLING SITES AND PROCEDURES:
# PRACTICAL THEMES FROM IRELAND

INGELISE STUIJTS

The Discovery Programme
34 Fitzwilliam Place, Dublin 2 (Ireland)
ingelise@discoveryprogramme.ie

ABSTRACT: This paper presents strategies for the sampling of charcoal and its subsequent preparation. The examples given are from recently excavated archaeological sites excavated in Ireland. The article follows the extraction process through to the identification procedures. Site conditions in Ireland vary, and both waterlogged and dryland sites are found. To date, much charcoal material recovered in Ireland is from excavations of *fulachta fiadh* or burnt mounds. These are often located on the margins between wet and dry terrain. Charcoal is also found in medieval corn drying kilns. Loose spreads of burnt material, ditch fills and (iron) smelting sites often form other rich sources of charred material. Much material investigated in Ireland is found in commercial excavations, where time and money constraints prevail. Nevertheless, working in this environment is not only challenging but also rewarding, not in the least because of the dynamics involved.
KEY WORDS: Charcoal, Flotation technique, Ireland, Fulachta fiadh

RÉSUMÉ: Cet article présente les protocoles de traitement des charbons de bois, depuis les stratégies d'échantillonnage jusqu'aux principes d'identification en passant par les techniques d'extraction du matériel. Les exemples donnés sont issus de fouilles récentes en Irlande où l'on retrouve aussi bien des sites d'ambiance humide que des sites terrestres secs. Beaucoup de matériel récupéré est issu d'un type de site très particulier à l'Irlande, appelé *fulachta fiadh*, et souvent localisé sur les marges entre terrain humide et sec. Les charbons proviennent également de fours médiévaux, de fosses ou encore de sites de fontes qui constituent des sources riches en matériel carbonisés. Enfin, il s'agit le plus souvent de fouilles préventives pour lesquelles le temps et l'argent sont une contrainte majeure. Cependant travailler dans ce contexte est à la fois un challenge et une récompense par la dynamique de travail engagée.
MOTS-CLÉS: charbons, flottation, Irlande, Fulachta fiadh

ZUSAMMENFASSUNG: Dieser Artikel beschreibt Strategien für das Sammeln und das anschliessende Aufbereiten von Holzkohle. Die gezeigten Beispiele stammen von kürzlich freigelegten Grabungsstätten in Irland. Der Artikel zeigt den Prozeß, die Holzkohle aus dem Fundstück zu separieren bis zum Identifikationsprozeß. Die Umstände der Ausgrabungsstätten in Irland variieren, es gibt sowohl nasse als auch trockene Stätten. Derzeit wird viel Holzkohle-Material in Ausgrabungen von *fulachta fiadh* (irish für: verbrannte Hügel) entdeckt. Diese finden sich häufig auf der Grenze zwischen nassem und trockenem Terrain. Holzkohle findet man auch in mitteralterlichen Getreide-Darrofen. Verstreutes verbranntes Material, grabenfüllungen und (Eisen-) Schmelzstellen bilden weitere reiche Quellen von verkohltem Material. Ein großer Teil des in Irland undersuchten Materials wird in kommerziellen Ausgrabungen gefunden, wo Zeit und Geld einschränkende Faktoren sind. Nichtsdestotrotz ist das Arbeiten in dieser Umgebung nicht nur herausfordernd, sondern auch belohnend, nicht zuletzt aufgrund der damit verbundenen Dynamik.
STICHWORTE : Holzkohle, Flotation, Irland, Fulachta fiadh

## INTRODUCTION

Charcoal is one of the most common environmental materials found in archaeological sites. The identification of charcoal is a common procedure following archaeological excavations. The material is chemically inert, which means that charcoal can survive a range of conditions, except for mechanical interference, such as smearing of wet charcoal through fingers.

Charcoal is the result of incomplete burning process,

either accidental or deliberate. The charcoal found on archaeological sites is often the result of firewood used in domestic fires. Studies of carbonized remains can provide evidence for the kind of wood usage, through, for example, the presence of charred chips or twigs.

In Ireland, the study of charcoal is usually carried out as part of the requirement to analyze material prior to radiocarbon dating. Before charcoal is analyzed, a license to alter from the National Museum is required. For the dating procedures a license to destroy/export is needed.

Due to pressures of commercialisation, not all the work that is currently undertaken by wood and charcoal specialists is being published. There are, however, some notable exceptions. Publications from research excavations include work by Sandra McKeown (1994) on material from the Bronze Age mines at Mount Gabriel and the Mesolithic material of Ferriter's Cove (1999). Other publications include those by Mark Hawthorne (1991) and Ellen O'Carroll (2001a, b). A recent publication from the private sector on excavations in Cashel, an urban settlement, is presented by Edmond O'Donovan et al. (2004). Wood and charcoal identifications from the Lisheen Mine Archaeological project in Co. Tipperary are published by Ingelise Stuijts (2005).

Despite the relative dearth of publications, commercial excavations also have their positive side. One factor that has not yet been fully appreciated is the enormous amount of archaeological material coming to light, excavated by licensed archaeologists.

Following road schemes and pipelines, large areas are monitored and subsequently excavated, giving an exponential rise in the amount of information available on particular regions. It is self-evident that the constraints of commercial excavations have an effect on the detail of information extracted from the ground. However, the excavation reports usually include several specialist reports, including charcoal analysis. It is an important task for the future to integrate this material into the wider archaeological discussion.

In this article, the practicalities of charcoal sampling of some typical (Irish) sites will be presented. This will be followed by a description of the most common features noted during the identification process. Most of the material described below derives from commercial excavations.

## ARCHAEOLOGICAL SITE SAMPLING

Charcoal remains can be found in a range of circumstances, both in waterlogged and dry deposits, and also sould be incorporated in minerals such as lime and iron.

It is common practise on most Irish excavations to take large bulk soil samples to accommodate a range of environmental studies. This prevents any bias by selecting

**Figure 1.** Killoran 10, a Middle Bronze Age cremation cemetery from the Lisheen Mine Archaeological Project
(Photo: Margaret Gowen & Co. Ltd., Dublin).

larger individual lumps or pieces of charcoal or other material. Bulk samples are then sub-sampled for both charcoal and macro-remains, such as seeds and fruits.

The bulk samples are collected in buckets of 5 or 10 litres, which are then sub-sampled. How much is sampled from each context depends on a number of (archaeological) factors, but often the quantities vary between 10 and 30 litres from each context.

Sampling is carried out in much the same way as for plant macrofossils, and normally all archaeological features are sampled. In large excavations, the area should be sampled at regular, square meter intervals using a grid system, although the exact density of the sampling would be discussed on site.

Features are sampled as a whole, except when they are very large and where sub-sampling is more practical. Large spaces such as floors or agricultural fields are sampled using a grid system; when layers are visible, they are sampled separately. Features such as wells and pits are sampled in horizontal layers. Cremations, urns and hearths are sampled in their entirely, whereas kilns are usually partially sampled. Figure 1 shows Killoran 10, a cremation cemetery dating to the Middle Bronze Age from the large excavations of the Lisheen Mine Archaeological Project (GOWEN ET AL. 2005). Hut remains from the same period were found in Killoran 8 (FIG. 2). The pits within these features were completely sampled and processed for charcoal analysis. Sub-samples were taken from the wall slots of the round houses.

## SAMPLE PROCESSING

Samples from dry-land archaeological sites that have little or no uncarbonized organic remains are processed using bucket flotation. This is a combination of flotation and sieving. The environmental material produced by this procedure is carbonised plant material such as charred seeds and charcoal. If samples are very large, a sub-sample of 5 litres is taken. If samples are smaller than 5 litres, the entire sample is sieved.

After recording the volume of the sample in a graduated bucket, the sample is saturated with water. Carbonised material will normally float to the surface. After agitating the sample the suspension is poured through a sieve with a mesh size of 250-300 μm. When no more material floats to the surface the remaining heavier residue that remained in the bucket is wet-sieved into a 1mm sieve. This procedure is carried out because not all carbonised material is light enough to float, and some may be left in the residues after flotation. Residues or retents may also contain other items of interest, such as small finds or non-carbonised environmental matter. After drying,

**Figure. 2.** Killoran 8, a Bronze Age round house from the Lisheen Mine Archaeological Project (Photo: Margaret Gowen & Co. Ltd., Dublin).

the smaller fractions of the flotation samples and the larger portions from the residues are kept separately into labelled plastic bags.

Charcoal can also be present in waterlogged situations. These samples are usually separated through a range of sieves, to collect the organic remains. For practical reasons, only the charcoal from the larger sieves is retrieved for identification. Lumps that are less than 0.5 $cm^3$ are usually not analyzed.

The samples are processed either on site (especially for the dry-land sites) or in laboratory situations (all wet sieving). The resulting charcoal is always thoroughly dried and left untouched and unsorted prior to handling by the environmentalist. If there is going to be a large volume of soil to float it is worth investing in a flotation machine, which allows for efficient processing of large volumes of bulk soil. For smaller excavations a simple flotation system using bucket and sieves would suffice.

## CHARCOAL IDENTIFICATION PROCEDURES

Microscopes are needed for the identification of charcoal. The practise of identifying charcoal in the field by using magnifiers of 10x should not be encouraged, as this magnification is not high enough to ensure a reliable identification.

Charcoal is broken along the three major axes (cross-section, radial and tangential) to expose a clear surface. After temporary mounting on clay or in sand, the surfaces are studied under indirect light. For species such as *Quercus* (oak) and *Fraxinus* (ash) the cross-section is often sufficient for a reliable identification. More characteristics are needed to identify other wood species, especially diffuse porous species such as *Alnus* (alder) and *Salix* (willow). Here, larger magnifications of 100-200x, sometimes 400x are used.

After identification, the separate species and categories are stored in labelled plastic bags. The weight of the sample and number of identified pieces are noted.

Bark pieces are not usually identified. Bark is also excluded from the identification list, although its presence, frequency and weight are noted.

There is some debate as to the number of individual lumps or pieces which should be identified to get a viable result. This discussion is often a consequence of financial constraints. The practise is usually to identify between 30 and 50 lumps per sample. This number is determined on the one hand by the actual number of charcoal lumps present within a cleaned sample, and on the other hand by the fact that many if not most species will be found within this number of identifications.

However, when scientific agendas allow more extensive analysis, a much larger number of identifications is preferable, sometimes more than 400 lumps per sample. Most species will generally be present in the first 30-50 lumps analyzed, although the rarer species are often those that give vital information on the local landscape in the past. In addition, when more observations are possible, a more confident conclusion can be reached. The observations may go beyond species identification and could involve branch sizes and annual ring patterns.

In samples with much oak charcoal, the aim to get 30 to 50 lumps per sample can be unsatisfactory, because oak can split easily and can dominate the sample. In these samples, priority is given to anything that is not oak. This involves scanning at lower magnifications and picking out the representative lumps. The latter method is especially useful for the investigation of Neolithic houses and medieval iron working sites. Here, the number of non-oak lumps is almost always less than 1%.

## CHARCOAL AS A SOURCE OF DETAILED INFORMATION

Depending on the quality of the charcoal, other observations can be made. These might include the presence or absence of bark, fungal hyphae (especially in the radial sections), insect channels and their size. To date there are no systematic measurements of individual charcoal lumps such as practised by Ludemann (2001).

In exceptional situations, beetles have been found within charcoal remains. The beetle remains are extracted and send for coleopteran analysis which can further enhance the environmental aspects of the study. The presence of beetles may indicate humid or wet conditions prior to burning, or simply the usage of rotten wood. The beetles themselves can carry information on the burning date, in that some adult beetles emerge at specific times of the year (REILLY 2005).

When complete twig or roundwood fragments are encountered, it can be noted whether growth is fast or slow, regular or irregular. A common feature is increasingly slower growth towards the bark, a fact that often is associated with a shrub-like growth and even the natural dying process of a branch or twig (SCHWEINGRUBER 2001). Fast-grown twigs may result from coppicing processes, although it is difficult to identify this from tiny charcoal lumps. When bark and pith are present, counting of annual rings and sometimes even the identification of the felling season is possible.

In charcoal research, measurements of annual rings are not usually made. It has, however, been observed that wide first and second annual rings occasionally occur, especially in charcoal from medieval samples. It is tempting to suggest that this might be the result of wood management.

## DENDROLOGICAL ANALYSIS OF CHARCOAL: SOME PROBLEMS

### Mineralization

Identification problems can arise in certain situations. In Ireland, one common environmental process is the incorporation of lime and iron into charcoal remains. This situation occurs especially in iron-rich clay soils but can also occur in cess-pits. Charcoal is either partly mineral replaced or can be coated with lime and iron minerals. These conditions can be found in the troughs of *fulachta fiadh* or burnt mounds. Similar difficult material was encountered by the author in the material from the Tell Ilipinar in Turkey (STUIJTS AND CASPARIE 1995).

The resulting charcoal can be as hard as stone and difficult to break. This has no consequences for radiocarbon dating, but the lumps are heavy compared to normal charcoal and hence more identifications of this material are needed for the dating procedures. Because of the relative weight, the charcoal fails to float and hence is only traced through sorting the residual fractions.

For the identification process the problem lies in the fact that the minerals have filled all cavities and that therefore the morphological characteristics are difficult to distinguish.

## Spiral thickenings

A peculiar phenomenon is the occurrence of features that look like spiral thickenings in the tracheids of *Pinus sylvestris* (Scots Pine), in particular. This may be a result of preservation conditions and is especially found in relatively old material such as Mesolithic hearths. It is important to concentrate on other characteristics to prevent identifications to e.g. *Taxus* (Yew), where spiral thickenings form one of the diagnostic features.

## ARCHAEOLOGICAL SITE VARIATION: SOME EXAMPLES FROM IRELAND

Where large-scale excavations bring occupation horizons or activity areas to light, a grid system for sampling is suggested. One such site is the Mesolithic settlement on Derragh Island, Co. Longford, currently being excavated by the Lake Settlement Project of the Discovery Programme (FIG. 3). This site lies adjacent to what was formerly an island in a large lake. The site occupies the fen margins and is very rich in organic remains, including charcoal and bone in excellent state of preservation. Because of the potential richness of information, the material was initially handpicked and every single piece measured. The site itself is excavated in a grid system with lines at 20 cm interval, and bulk samples were taken from every square metre (FIG. 4). This method is here especially useful, because the site seems undisturbed and has at least two phases of activity,

separated by a distinct layer of peat (FREDENGREN *ET AL.* FORTHCOMING).

The grid system of charcoal sampling has been used in the past in Ireland by McKeown (1994, 1999). Her investigations provided extensive information on the prehistoric mining activities in Mount Gabriel, Co. Cork and in the cave deposits of Ferriter's Cove, Co. Kerry.

One of the most common archaeological sites in Ireland that are excavated every year are *fulachta fiadh* or burnt mounds (FIG. 5). Current research suggests that they are a typical element of Irish Bronze Age settlement architecture (BRINDLEY *ET AL.* 1989/90). They are characterized by dumps of burnt stones and charcoal around a water-filled pit or trough (FIG. 6). The water in the through was heated for a variety of domestic functions including cooking and bathing by immersing heated stones. They are the most common prehistoric monument type in Ireland, alongside early medieval ringforts.

*Fulachta fiadh* are associated with kidney- or horseshoe-shaped mounds of charcoal and heat-shattered sandstone or limestone, although when excavated, sometimes only a spread of burnt stone is all that survives. They are usually associated with hearths where the stones were heated. The mound is the result of accumulating debris from the repeated use of the stones to heat the water. They are commonly located close to a water source or in marshy ground so that the pit quickly filled with water. The function of the pit was to collect water and sometimes the

**Figure 3.** Derragh Island in Lough Kinale, Co. Longford (Photo: The Discovery Programme).

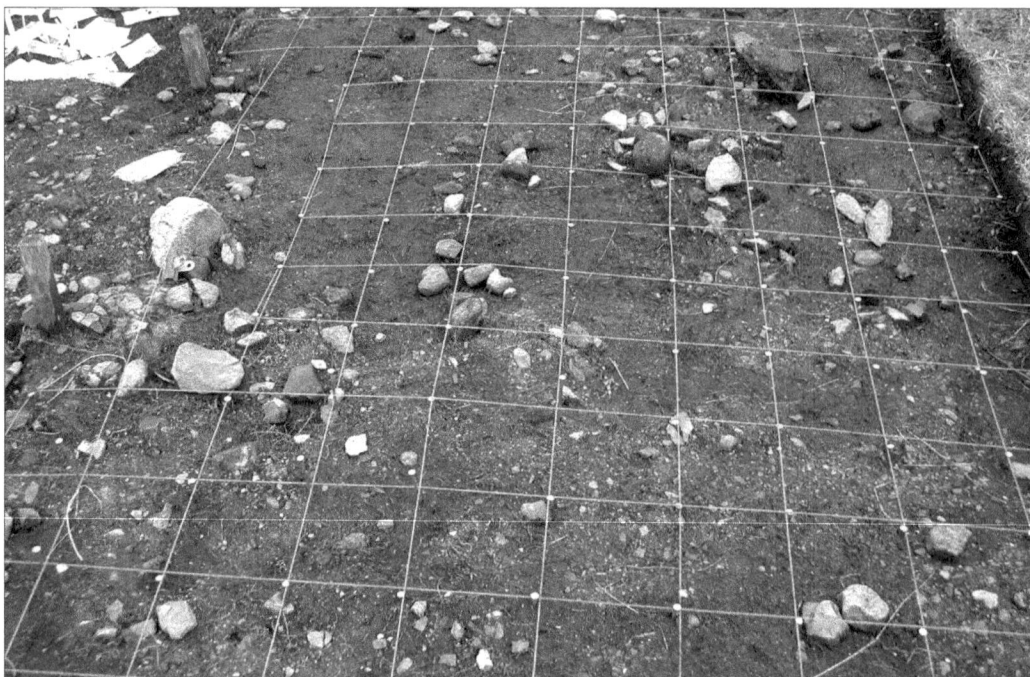

**Figure 4.** Derragh Island: excavating using a grid system with lines at 20 cm interval (Photo: The Discovery Programme).

**Figure 5.** Attireesh 3 *fulacht fiadh*: mound in the field (Photo: Richard Gillespie, Mayo County Council).

remains of a wooden or stone lining is found, into which the heated stones were thrown (FIG. 7 and 8).

*Fulachta fiadh* have been interpreted as cooking places although few finds are associated with the excavated examples and bones are not often found. Experiments by O'Kelly (1954) have shown that a through of water could be brought to the boil using heated sandstone in just 30-35 minutes. In his experiments, food was wrapped in straw and put into the boiling water. Only a few heated stones were required to maintain the temperatures. The amount of fire-cracked stones indicated that the sites were used on

**Figure 6.** Attireesh 2. trough with fill intact (Photo: Richard Gillespie, Mayo County Council).

**Figure 7.** Killoran 253 wicker-lined trough, Late Bronze Age (Photo: Margaret Gowen & Co. Ltd., Dublin).

numerous occasions. Other researchers suggest a use as a washing or bathing place, a sauna bath or a site for textile production or felting wool. It has been proposed (DOODY 1993), based on their relatively restricted date, that their very numbers could indicate transitory settlement perhaps associated with transhumance or "booleying".

The quantity of burnt stone and charcoal associated with *fulachta fiadh* suggests a certain period of usage. It is likely that the troughs would have to be cleaned at regular intervals, prior to the next use, and the cleared material was thrown aside, forming the mound. Interestingly, in charcoal samples from several areas within the same

**Figure 8.** Killoran 265 hollowed-out oak (Photo: Margaret Gowen & Co. Ltd., Dublin).

wood suitable for firing purposes. Local felling of trees for timber, for example, would leave a large proportion of the tree available for firewood. If convenient, waste material such as chips or discarded building material or woodworking waste could also be used. The composition of charcoal assemblages might thus be influenced by other patterns of wood use.

Especially in historical times, part of the local woodland was managed specifically to provide wood fuel or material for charcoal burning for metalworking. Oak and ash both produce good fuel and are also excellent trees for timber. It is likely therefore, that ash and oak would have been encouraged in managed woodland. In this light it would be interesting to compare the charcoal content of corn-drying kilns and possible structural elements within those kilns.

mound at Attireesh, Co. Mayo, the species represented and their proportions were identical (Attireesh area 1, in STUIJTS FORTHCOMING). However, more data are needed to ascertain whether this idea is as a result of disturbance to the mound, or whether there was a consistent use of certain species.

Charcoal is the most common ecofact found in excavations of *fulachta fiadh*. It may be present in large quantities, especially when remains of wooden troughs are present. Because of the quantity of charcoal, burnt mounds features are not sampled as a whole. It is generally recommended that two samples are taken from the trough (bottom and top) and one from the mound. The charcoal is used for dating purposes, and also provides insight in the local vegetation around the site at the time of use.

Corn-drying kilns are regularly encountered in excavations from the Early Medieval or later periods in Ireland. Here, large quantities of burnt cereals and charcoal are present. Usually only part of the material is sampled from kilns, because of the amount of material present.

In Ireland there are relatively few wood species that occur naturally in the prehistoric period when compared with assemblages from mainland Europe. In historic times, with imported wood species and plantation, a more diverse pattern can be expected. In Ireland this often means the presence of conifers such as *Picea (*spruce) and *Pinus sylvestris* (Scots pine).

It has been demonstrated that commercial excavations in Ireland are very productive. *Fulachta fiadh* or burnt mounds have been mentioned as one of the most common archaeological sites currently under investigation. Charcoal from excavations must be identified prior to dating procedures. This fact has led to an increase in the quantity of material being analyzed and hence the information being collected. The procedures for sampling and extracting charcoal follow those for other ecofacts especially macro-remains or seeds.

## INTERPRETATION OF THE RESULTS AND CONCLUSIONS

It is usually assumed that the firewood was collected as close as possible to hearths, so that the wood species found here would therefore provide information on the local vegetation. Other activities, however, could also produce

ACKNOWLEDGMENTS

The author would like to thank the Discovery Programme, especially Dr. Annaba Kilfeather and Anthony Corns, for their help with the practicalities of this publication. For the material and pictures thanks goes to Dr. Christina Fredengren of the Discovery Programme, to Margaret Gowen & Co. and the Lisheen Mine Archaeological Project, and to Richard Gillespie, archaeologist for County Mayo. Birthe Stuijts and Dr. Alexa Dufraisse made the German respectively French summary.

REFERENCES

BRINDLEY A.L., LANTING J.N., MOOK W.G., 1989/1990.- Radiocarbon dates from Irish fulachta fiadh and other burnt mounds, *Journal of Irish Archaeology*, 5: 25-33.

DOODY M.G., 1993.- Bronze Age Settlement, *in*: E.S. Twohig and M. Ronayne (eds.), *Past Perceptions: The Prehistoric Archaeology of South-West Ireland*, Cork, Cork University Press: 93-100.

FREDENGREN C., KILFEATHER A., STUIJTS I., FORTHCOMING.- *Lough Kinale*.

GOWEN M., Ó NÉILL J., PHILLIPS M., 2005.- *The Lisheen Mine Archaeological Project 1996/8*, Bray, Wordwell, 387 p.

HAWTHORNE M., 1991.- A preliminary analysis of wood remains from Haughey's Fort, *Emania*, 8: 34-38.

McKEOWN S., 1994.- "Wood remains", *in*: W. O'Brien (ed.), *Mount Gabriel: Bronze Age Mining in Ireland*, Galway, Galway University Press: 265-281.

McKEOWN S., 1999.- Charred wood, *in*: P. Woodman, E. Andersen and N. Finlay (eds.), *Excavations at Ferriter's Cove, 1983-95: last foragers, first farmers in the Dingle Peninsula*, Bray, Wordwell: 213-217.

LUDEMANN T., 2001.- Das Waldbild des Hohen Schwarzwaldes im Mittelalter, Ergebnisse neuer holzkohleanalytischer und vegetationskundlicher Untersuchungen, *Alemannisches Jahrbuch*, 1999/2000: 43-64.

O'CARROLL E., 2001a.- Analysis of archaeological wood found in Irish bogs, *in*: B. Raftery and J. Hickey (eds), *Recent Developments in Wetland Research*, Dublin, Seandálaíocht: 27-35 (Mon. 2. Department of Archaeology, University College Dublin and WARP Occasional Paper 14).

O'CARROLL E. 2001b.- *The Archaeology Of Lemanaghan. The Story Of An Irish Bog*, Bray, Wordwell.

O'DONOVAN E. *ET AL*. 2004.- Excavations at Friar Street, Cashel: a story of urban settlement, *Tipperary Historical Journal*, 2004: 3-90.

O'KELLY M.J., 1954.- Excavations and experiments in ancient Irish cooking-places, *Journal of the Royal society of Antiquaries of Ireland*, 84: 105-155.

REILLY E. 2005. Coleoptera, *in*: M. Gowen, J. Ó Néill, and M. Phillips (eds.), *The Lisheen Mine Archaeological Project 1996/8*, Bray, Wordwell,: 187-208.

SCHWEINGRUBER F.H., 2001.- *Dendroökologische Holzanatomie. Anatomische Grundlagen der Dendrochronologie*, Bern/Stuttgart/Wien, Verlag Paul Haupt, 472 p.

STUIJTS I.L.M., CASPARIE W.A., 1995.- Ilipinar wood remains, *in*: J. Roodenberg (ed.), *The Ilipinar Excavations I, five seasons of fieldwork in NW Anatolia*, 10. Nederlands Historisch-Archaeologisch Institut: 157-158.

STUIJTS I. FORTHCOMING.- Wood and charcoal identifications. Westport main drainage and waste water disposal scheme contract 4, County Mayo.

STUIJTS I. 2005.- Wood and charcoal identification, *in*: M. Gowen, J. Ó Néill and M. Phillips (eds.), *The Lisheen Mine Archaeological Project 1996/8*, Bray, Wordwell: 137-185.

# MINE CHARCOAL DEPOSITS: METHODS AND STRATEGIES.

# THE MEDIEVAL FOURNEL SILVER MINES IN

# THE HAUTES-ALPES (FRANCE)

Vanessa PY

MMSH, Laboratoire d'Archéologie Médiévale Méditerranéenne, UMR 6572 CNRS
Université de Provence - Aix-Marseille I
5 rue du Château de l'Horloge B.P. 647 F-13094 Aix-en-Provence cedex 2 (France)
py@mmsh.univ-aix.fr

**ABSTRACT:** The medieval Fournel silver mines in the Hautes-Alpes (France) are an original archaeological cadre for anthracological study. Massive use of fire for breaking down the hard quartzite bedrock developed working strategies which conditioned the architecture of the workings, operational dynamics, morphology of wastes and their management (storing, backfill). These wastes hold large quantities of charcoal, the ultimate traces of the thousands of fires which opened up exploitation of the silver galena (lead sulphur with 0.15% silver). The awkwardness of the archaeological context and the size of backfill conditioned creation of an adapted sampling protocol. Study of these deposits, combined with archaeological analysis of the workings and a sedimentological approach to backfill, has lead to reconstitution of the area where fuel was obtained for the mines. First results allow interpretation in fuel management and environmental adaptation. This communication examines the preliminary studies carried out during a Master's degree thesis on forests and proto-industry in the High Durance valley during the Middle Ages and during scheduled archaeological excavations.
KEY WORDS: mining, fire-setting, techniques, fuel management, charcoal, anatomical signatures

**RÉSUMÉ:** Les mines médiévales du Fournel dans les Hautes-Alpes (France) constituent un cadre archéologique original pour une étude anthracologique. L'utilisation massive du feu pour attaquer la roche encaissante particulièrement dure (quartzites) a généré des stratégies d'exploitation qui conditionnent l'architecture des ouvrages, la dynamique opératoire, la morphologie des déblais et leur mode de gestion (stockage, remblaiement). Ces résidus renferment de grandes quantités de charbons de bois, ultimes traces des milliers de brasiers qui ont permis d'exploiter la galène argentifère (sulfure de plomb titrant 0,15 % d'argent). L'exiguïté du contexte archéologique et l'ampleur des remblais de taille au feu ont nécessité la mise en place d'un protocole de prélèvement adapté. L'étude de ces dépôts, combinée avec l'analyse archéologique des ouvrages et avec l'approche sédimentologique des remblais, permet de reconstituer le territoire d'approvisionnement en bois de feu des mineurs. Les premiers diagrammes autorisent des interprétations en termes de gestion et de modes d'adaptation aux disponibilités environnementales. Cette contribution fait état de travaux préliminaires menés dans le cadre un Diplôme d'Etudes Approfondies sur les protoindustries et la forêt dans la haute vallée de la Durance au Moyen Âge et d'une fouille archéologique programmée.
MOTS-CLÉS: mine, Moyen Age, abattage au feu, charbons résiduels, représentativité, paléoécologie

**ZUSAMMENFASSUNG:** Die mittelalterlichen Bergwerksanlagen des Fournel im Département Hautes-Alpes (France) bilden einen ungewöhnlichen Rahmen für eine anthrakologische Studie. Die ausgiebige Anwendung von Feuer zur Sprengung eines besonders harten mineralhaltigen Gesteins ging mit Abbaustrategien einher, die den Aufbau der Holzschichtungen, die Verfahrensweise, die Morphologie des Abraums und deren Entsorgung (Halde, Zuschüttung) beeinflußten. Die Brandrückstände enthalten große Mengen von Holzkohle, Reste tausender Brandherde, die zum Abbau des silberhaltigen Bleierzes (zu 0.15% silberhaltiger Bleiglanz) eingesetzt wurden. Die beschränkten Möglichkeiten der archäologischen Untersuchung und der Umfang der durch die Feuersetzung angefallenen Rückstände machten eine methodische Absicherung der Probenentnahme erforderlich. Die Analyse der Rückstände im Verband mit der archäologischen Untersuchung der Holzaufbauten und der sedimentologischen Erfassung des Abraums ermöglicht es, das geographische Umfeld der Holzbeschaffung durch die Bergleute für die Feuersetzungen zu rekonstruieren. Die ersten Diagramme lassen Deutungen zum Rohstoffhaushalt und zur Anpassung an die Umweltbedingungen zu. Unser Beitrag stellt erste Ergebnisse von Vorarbeiten vor, die im Rahmen eines Dissertationsvorhabens über die frühindustrielle Nutzung des Waldes im oberen Durancetal im Mittelalter mit einem Grabungsprogramm durchgeführt wurden.
STICHWORTE: Bergbau, Mittelalter, Feuersetzen, Brandrückstände, Holzkohle, Paläoökologie, exemplarische Studie

# INTRODUCTION

Before the invention of "modern" fuel, the use of explosives and later material for underground planning, (cast iron canalisation, railway tracks for example), an almost organic relation existed between the mine and the forest. All activities leading to the operating chain of metal production, consume wood. One after the other or simultaneously, they are at one moment or another of history, accused of being responsible for enormous destruction of forests and are therefore subjected to strict regulation. The ancient mining method of firesetting has long been the source of pessimistic descriptions concerning precocious deforestation near a mine site. During the nineteenth century the deforestation of mountains is partly imputed to this method called archaic, but still largely practised at that time in the Hartz mines in Germany and also at Kongsberg in Norway (BERG 1992). These regions are in fact well known for their inexhaustible forest resources, but are they not guardians of a rather efficent replanting policy? Actually, no reliable figures are available to feed this argument and show the theory of a massive and irreversible "mining deforestation". In spite of these gaps, we are obliged to consider the relation that exists between certain extraction sites and the particularly denuded areas such as the Brandes plateau in the Oisans region (French Alps).

The abundance of charcoal conserved in the underground backfill and in the waste heaps of the Fournel mines (L'Argentière-la-Bessée) authorises a profond charcoal study. They synthesise an intensive consommation of wood that stretches over at least four centuries. Thus it is possible to detect in first anthracological diagrams a behavioural evolution of supplies in firewood and to make an estimation of the available ligneous biomass. However, to ensure operational lines of work it has been necessary to develop methods and strategies proper to this original archaeological programme. A notable problem was the paleological representativeness of charcoal issuing from an extraction technology.

## STUDY FRAMEWORK

### Geography, Biogeography and Geology

One part of the ancient mining works lies at the bottom of the enclosed valley of the Fournel which forms a gorge 1 km in length. Upstream, it emerges into the glacial valley of the Alp Martin situated on the Eastern face of the Ecrin massif between the intermediary and internal Alps. Its catchment area belongs to the hydrographic basin of the High Durance. Downstream, the gorges open progressively and emerge into the Durance valley (FIG. 1). Above the gorges, on a steep wooded versant,

situated on the right bank, lie some ancient crumbling works. On the right bank, a very steep slope, prolonged by an incline more regular and occupied by the hamlet of "l'Eyssaillon" at 1100 m of altitude. It continues thus until the summit of the "Têtes" which culminates at 2000 m of altitude. Downstream, this incline is brusquely interrupted by a rocky shoulder of quartzites, lightly cut by several thalwegs. The ancient works and modern research are situated in a level variation of 800 m (FIG. 2) (ANCEL ET AL. IN PRESS a).

This region is characterised by terraced vegetation and a strong contrast "ubac"/"adret". In the "adret", the bottom of the versant is characterized by grassy pre-steppic levels of *Juniperus* and *Lavandula* (900-1200 m alt.). A rich shrubby vegetation is developing in rough land at the edges of meadows and terraces (*Berberis vulgaris, Prunus spinosa, Cytisus, Cornus sanguinea, Crataegus, Fraxinus excelsior*). Forests of hardwood and shrubs grow at the bottom of the versants of the "ubac" and on the outskirts of the Fournel (*Corylus avellana, Sorbus aucuparia, Populus tremula, Fraxinus excelsior*). Then the mountain shelves stretch up to 1500-1800 m of altitude. The "adret" is characterized by a heliophilous forest of *Pinus sylvestris*, the heathland with *Juniperus* in the warm and sunny parts, and hardwoods at the beginning of the valley (*Fraxinus excelsior, Acer pseudoplatanus, Rosa*). In the "ubac", the extension of the pinewoods is limited. The ancient pastureland is rapidly surmounted by larch plantations and thickets of *Sorbus aucuparia*. In the valley, sub-mediterranean influences raise the limits of the mountain shelf to the "ubac" around 1600 m high and at the "adret" up to 1800 m. The humid forests exposed to the North are composed of fir plantations of *Abies alba* mixed *Larix decidua*. The subalpine shelf reaches up to 2000-2200 m in height. It is notable essentially for its larch plantations in the meadows at the lower limit and the rhodoraie on the upper limit (MEYER 1981, RAMEAU ET AL. 1993).

From a geological point of view, the mine is situated in the Briançonnais zone of the Alpine massif. From East to West, the ground is schematically composed of carboniferous sandstone, of permian conglomerates (Verrucano facies), of quartzites and of triassic limestone. They have been uplifted vertically by the thrust and are dislocated by numerous faults. There exist several seams of the Eocene to Oligocene periods, enclosed in quartzites. They are cut in panels orientated North-East/South-West, sub-vertical or inclined toward the South-East, at 15 to 50°. Their length varies from 0.50 m up to 7 or 8 m. The predominant ore is argentiferous galena. The matrix is composed of quartz and barytine. This ore is more or less concentrated. The richest part shows between 10 and 30%. It was mined exclusively during the Medieval epoch then privileged by the Moderns (ANCEL ET AL. IN PRESS a).

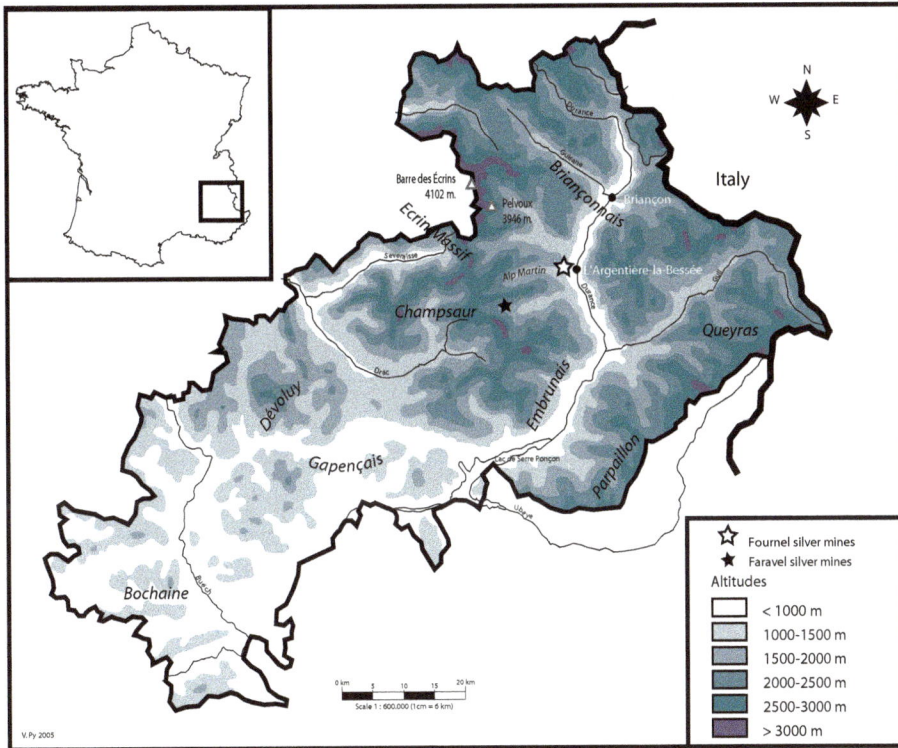

**Figure 1.** Localisation of the Fournel and Faravel mines (Hautes-Alpes, France).

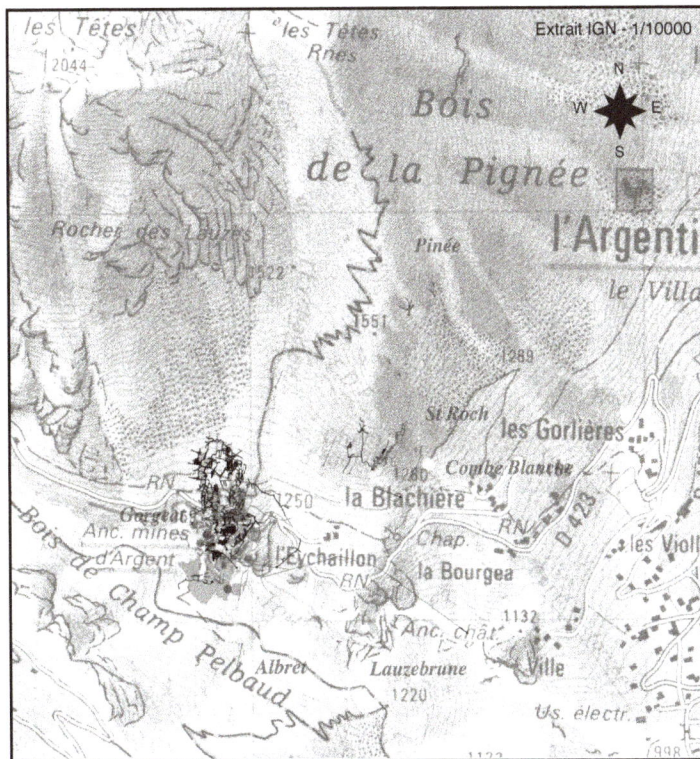

**Figure 2.** Localisation of the medieval workings (Extrait IGN 1/10000, B. Ancel 2004).

## Archaeological context

The prospections[1] conducted on the rock outcrops show numerous traces of ancient workings began as open cast exploitations for over 4000 m² (Ancel 1998a). The medieval workings are situated in the Fournel gorges, in the localities of "Gorgeat" and "Lauzebrune" and, on the "*adret*" slope, at "Saint Roch", "Comble Blanche", the "Rouille" and the "Pinée". They are composed of small sized galleries and firesetting sites mostly backfilled. They form five principal sites quite distinct, which reach in to over 100 m from the face (Fig. 2). According to approximate estimations, the miners deplaced a total of 20000 m² of mineralisation. This is a considerable volume to have excavated.

Firesetting was initially used to mine quartzites, the surrounding rock being for the most part very hard. The use of this technique is characterised archaeologically by rounded vaults covered in soot and by rich charcoal deposits in the backfill (Ancel *et al.* in press b). Owing to the size and good state of conservation of the site, the archaeologists were able to elaborate interpretations of the operational dynamics and characterise the general organisation of the underground site. The steepness of the place did not constitute a real handicap for this exploitation based on manual work and carriage on the backs of men and mules. The strong level variation facilitated vertical development of the sites, evacuation of used air by thermic aspiration and drainage of infiltrated water by gravitational flow. The sites marry well the geometry of the stratum with the works assistance, developed as systematic control of problems (Ancel 1998a, b, 2000).

## METHODS AND STRATEGIES: SEVERAL VIEWPOINTS

The exiguity of the medieval workings (caving access), the stratigraphic complexity of backfills and the relatively unknown uses of wood in firesetting, necessitated putting into place a pluridisciplinary approach. Study of the combustible requires an experimental approach to characterise notably properties of the wood, its burning qualities, combustion process in a semi-closed situation (air currents) and charcoal fragmentation. Their macro- and microscopic analysis permitted to build up the basis of a collection of references on the state of the wood and the traces of burning[2]. Concurrently, field study coupled with a sedimentologic approach is realised on the backfill, product of the accumulation of poor or sterile wastes, issued directly or indirectly from firesetting.

## Archaeological approach

For the last twenty years, anthracological work has shown the major importance of sampling archaeological charcoal to find information exhaustively (quantitative) and obtain a representative corpus (qualitative) (Chabal 1982, 1988, 1991, 1997, Heinz 1988, Badal Garcia 1990). The method can be applied to the mining context following particular modalities.

In the zones near the open cast sites, the ancient backfill is masked by gravel more or less infiltrated by sand mixed with earth. Access can require an important work of cleaning and clearing. On top of this, these zones have been affected by exploitation attempts in the nineteenth century. The wastes mostly untouched are often lying in remote and inaccessible parts of the mine. They have been stocked by the miners, who reserved circulation passages and, abandoned once the vein petred out. This rigorous method of deads management avoided the long and difficult work of carriage to the outside. To obtain deposits best characterizing the extraction of a cavity, the works where the access is delicate are privileged. They exclude the systematic evacuation of wastes towards the exterior.

Excepting the mining heads, a plotting dig with a squaring method would not be considered underground. The charcoal deposits are sealed in beds composed of gravels, sand and centimetric to decimetric blocks, too unstable for excavation. The adopted strategy consists of effecting successive stratigraphic sections in the trenches (opencast) or the access works or, enormous transversal cuts of the whole site. They are photographed, plotted and described in detail according to their granulometry, the inclusions and their colour.

This method brings out anthracological samples in every characterized layer, renewed in every successive cut (1 to 2 m of interval) or on the whole length of the big transversal cuts. Spatial analysis of the charcoal deposits is therefore possible. From 10 to 30 litres of firesetting waste can be taken from each layer, depending on their richness. This method permits acquisition of a rigorous sample, representative of the statistic population of a deposit. The difficulties of moving a load imposes sifting *in situ* where floatation is excluded. The humidity of the waste necessitates the uses of a column of sifter screens 8-4 mm to evacuate the bigger elements and avoid obstruction of the finer mesh. The material thus sifted is integrally recovered. It is dried then classed in a laboratory. A second sifting using water can be necessary for the dirtier wastes. The charcoals smaller than 4 mm are excluded from the study because they come from the fragmentation of the bigger pieces. The fragments bigger than 4 mm offer the same proportion between types than the charcoals composed between 0.5 and 3 mm and their floristic content does not seem different (Badal-Garcia 1990, Chabal 1997: 37).

A paleoecological and ethnobotanical approach must be envisaged in the long term to perceive the significant changes and the installation of techniques. The anthracological sampling must be realised on the scale of the site (gallery, hall) but also to the scale of the network. This imperative implies a tight cooperation with mining archaeology which seeks to determine a relative chronology with the works. Several are refilled with tons of waste and disfigured by the resumption of work in the nineteenth century. Thus, the interpretation of the operating dynamic is incomplete. Considering the scale of the site, the modalities of management of wastes as the mining advanced are difficult to perceive (ANCEL IN PRESS). Radiocarbon dating proposes a bracket still too large and time passed between each deposit is hypothetic. A multiplication of radiocarbon dating within the same working and their comparison with sedimentological statistics, could furnish the key to the interpretation.

## Sedimentology approach

To apprehend precisely the operating dynamic and management of the steriles by the miners, Christophe Marconnet proposes a new method based on protocols of sedimentology[3]. The sand and bigger elements are sampled in stratigraphic sections and are subjected to granulometric classement and macroscopic analysis. A similar study is conducted on the residues set apart in the materials obtained during firesetting experiments (Fournel mine). This permitted to identify granulometric characters proper to each stage of the firesetting and the factors susceptible to modification (MARCONNET 1994). The results of analysis accomplished on the archaeological samples are more heterogeneous and diversified. They are without a doubt allied to the firesettings realised in varying rock contexts. But this variety of granulometric facies underlines especially the existence of an operating model more complex than the experimental protocol, comprising the pick up, transport, stockage and sorting stages, evidenced by the stratigraphy. The comparative study of these residus of different origins, permits progression in the comprehension of the firesetting techniques.

## Anthracological approach

### Underground wastes

The layers of wastes containing charcoal inclusions interpolate themselves with "clear" layers. The charcoals are found either in a dispersed form with small sized fragments (4-6 mm) hardly visible to the naked eye and spread in a homogeneous way in the layer, or under a concentrated form with bigger centrimetric fragments. Easily accessible, they were selected essentially by archaeologists to meet the quantity necessary for radiocarbon analysis. They were subjected to sampling to date the exploitation of different networks.

In spite of the "choice" induced by the selection of the bigger elements to the detriment of the finer fractions, these charcoals were subjected to a first experimental analysis (sample portion of type 1). The total being, 1079 fragments selected in the areas of "Vieux Travaux" (Gorgeat), of "Saint Roch", of "Lauzebrune", of the "Pinée" and the "Combe Blanche" have revealed three different taxa: *Larix-Picea* (*cf. Larix*), *Pinus* type *P. sylvestris* and *Abies* sp. *Larix-Picea* appear to be used in priority. Nevertheless, some remarkable variations appear as in the diagram of "Saint Roch" where *Abies* sp. shows 60.5% of relative frequences against 39.5% for *Larix-Picea* and, in the case of "Vieux Travaux" where *Pinus* type *P. sylvestris* is dominant in two sequences of the diagram (FIG. 3). In the absence of a preliminary methodological reflection, these inversions of the dominant taxa (or secondary in the case of *Abies* sp.) could be interpreted as a change in the wood supply modalities, but, they are probably the truncated image of backfill of odd logs or the spreading of centres of woodcutting rarely touched. A second campaign of sampling was therefore begun following the modalities presented earlier and targetting the dispersed stratified deposits which could offer a good synthetic effect of the combustible used.

A total of over 1500 new fragments have been subjected to preliminary anthracological study. They revealed very clear refinement of proportions between the majority of taxa and a light opening of a floristic spectrum with the appearance of hardwoods of which the proportions were anecdotal (FIG. 4). *Pinus* type *P. sylvestris* poorly represented or over represented in the samplings of "type 1", constitute in the samples of "type 2", 1/10 to 1/3 of the combustible put to work for the firesetting. Under this title, the sequence of "Lauzebrune", dated eleventh century, is very significant. In the sample E28 (type 1), *Larix-Picea* totals 100% of the relative frequencies whereas in the samples 2P2 and 1P1 (type 2) *Pinus* type *P. sylvestris* reach 22.6% and new taxa appear: *Fraxinus* sp. and *Acer* sp.

In the locality of "Vieux Travaux", a "test" sample realised from a stratigraphic section in a waste heap offered a frequency distribution quite homogeneous in the four identified layers. However, the frequency intervals between the different taxa do not overlap. A $X^2$ test realised from the relative frequencies of different taxa in each sample show a highly significant difference of the distribution of frequencies in the four layers. According to the calculations of contributions *a posteriori*, the samples 17 and 19 are responsible for the high value of the $X^2$ test. It seems that the number of fragments analysed (50) in the sample 17 have influenced the proportions between the *Pinus* type *P. sylvestris* and *larix-Picea*. In return, the sample 19 counts 100 fragments, a consequent amount compared to the little portion of sampled layers. This established fact leads to the problem of the optimal amount necessary in each layer.

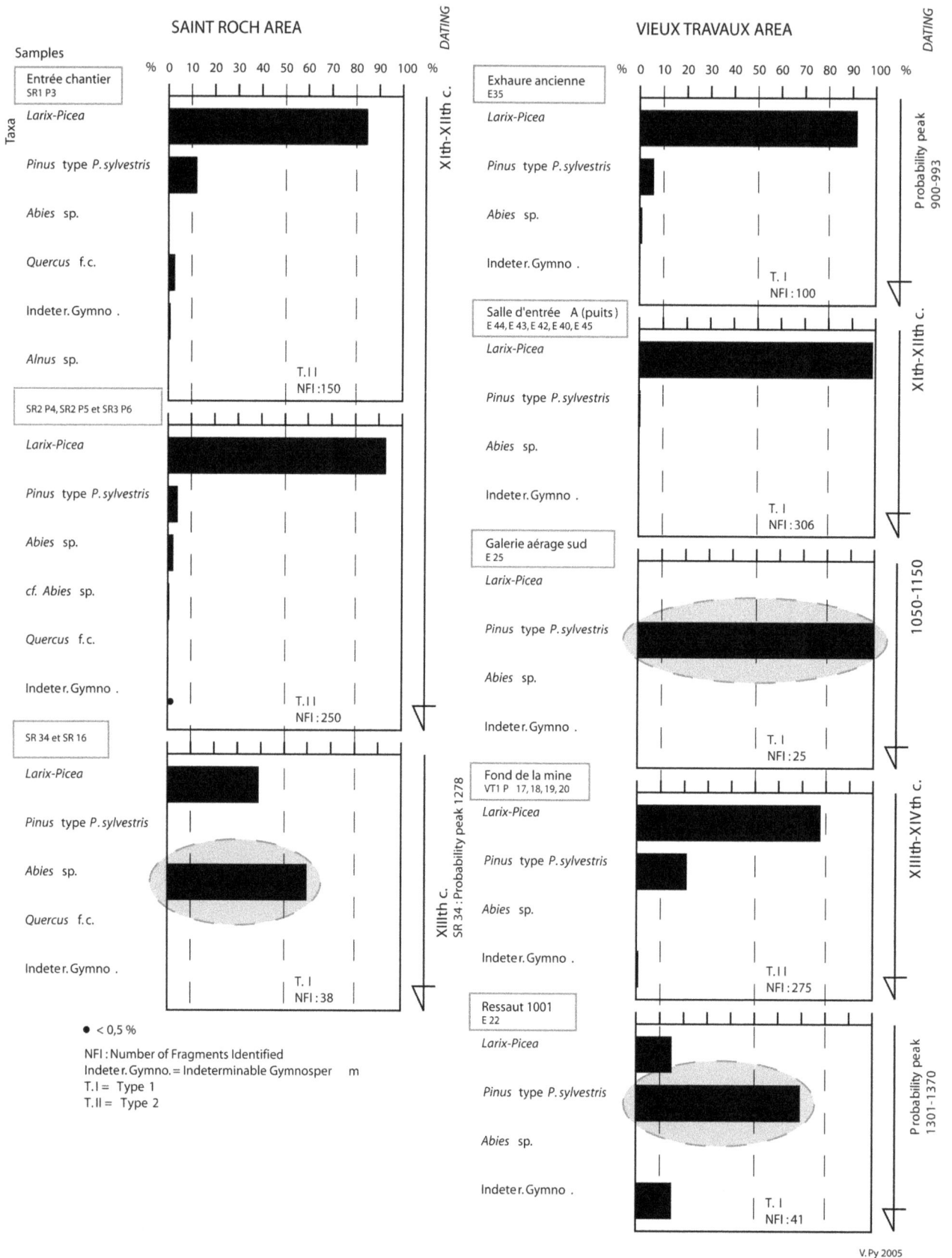

**Figure 3.** Significant inversions of the dominant taxa in samples "type 1": Example of anthracological sequences of "Saint Roch" and "Vieux Travaux".

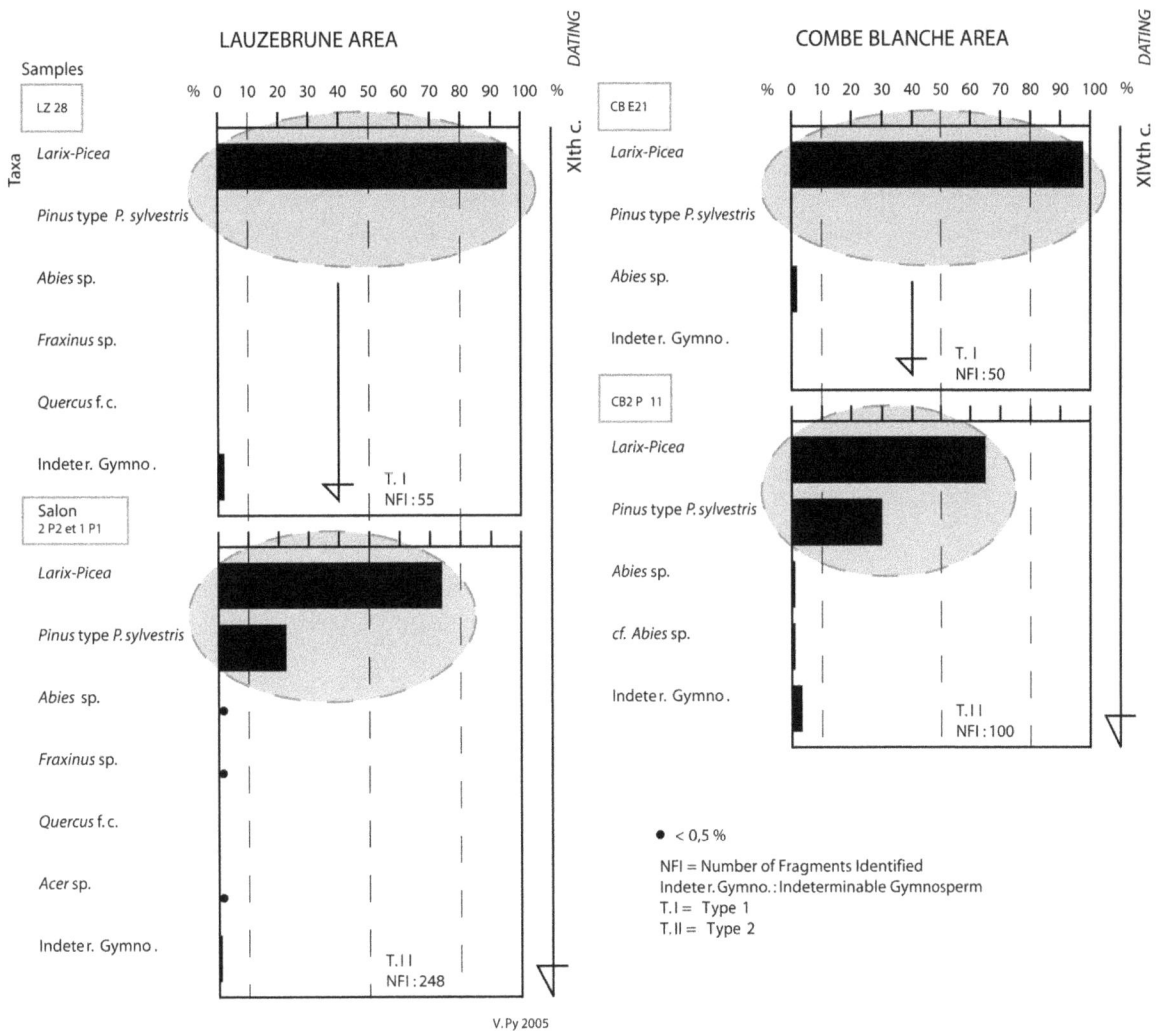

**Figure 4.** Enlargement of the spectre and refinement of the proportions between the dominant taxa with type 2 sampling. Examples from the anthracological sequences from Lauzebrune and Combe Blanche sectors.

The number of taxa potentially present is too reduced in this type of deposit to have recourse to the effort-yield stage. With a random sampling of the fragments, a frequent variety has more chance to surface than a rarer one. It is therefore important to sample a consequent amount to come up with a less common variety and hope to obtain refined proportions between taxa. In this specific case, a minimum amount of 50 to 100 charcoals each layer in a same stratigraphic section is therefore largely accurate.

*The wasteheap potential*

The totality of medieval exploitation remains, visible on the surface, is tied to the extraction of ore. It concerns the open cast sites, the orifices of tunnels and the vast pile of waste characteristic of a mining scheme: the heaps. They are still well apparent on the slopes of "Saint Roch", of the "Rouille" and of the "Pinée", but their morphology is considerably altered by reason of erosion (ANCEL 1998a, b).

West of "Combe Blanche", extends a vast modern heap composed of very fine waste of whitish colour, characteristic of quartzites and barytine. They cover older rejects and mask the presumed entrances of "Saint Roch". On the Eastern versant which dominates the town of L'Argentière, the quartzites are outcropped on the scree summits. The "rust" colour which affects certain flows signals the presence of rubified matrix. On the south side, the most important flow is topped by a plateau of waste (large shoulder). To the north of "Rouille", on the right hard side of the big screes, the principal vein has been subjected to an open caste scrape on 1600 m². The summit of the layer is recut by a mineralised fault which has been partly exploited by fire. Important quantities of residual charcoal are spread on the ground. In the gorges, the rejects of deads carried by colluvial action and by flooding of the torrents have practically disappeared.

These important spreads of sand and gravel are formed

by the accumulation of successive scrapings of the rock. They do not contain all the wastes of the miners who searched obstinately to stock them in the mine to reduce to the maximum of the work of evacuating them to the surface. The compiled analysis of the surface waste and the underground backfill is therefore complimentary. It seems pertinent to evaluate the statistical frequence of charcoals contained in these deposits and to test their anthracological potentiel. The stratigraphy corresponds in shape to the dip of the versant and is composed of alternate layers of sandy matrix, including gravel and small stones of centimetric to decimetric size and more rarely of blocks.

The samples taken in layers 2 and 3 of the sampling opened in the "white heap" gave good results with 30 litres of waste per layer. In total, the analysis of more than 600 fragments offers good broadening of floristic spectrum with the appearance of new taxa such as *Juniperus* sp., *Laburnum-Cytisus* and *Ulmus* sp. (FIG. 5). The degree of effort-production tends to stabilise beginning at 250 charcoals analysed. One approaches the normal standards of the classical archaeological context (CHABAL 1997).

A stratigraphic section of 3 m in length, realised in the surface waste at the entry of the "Pinée" network, permitted a second sample "test" in a coal layer sealed between the "clean" layers. 40 litres of waste sampled gave 265 charcoals. In this case, the degree effort-production tends to stabilise at the 150-200 level of charcoal analysed.

The minimum amount of charcoals to sample in the deposits of "wasteheap type" can be evaluated at least 200 fragments by layer to attain the level. Anyway, as several species totalise almost all the frequencies (*Larix-Picea* et *Pinus* type *P. sylvestris*), a superior number can be necessary for a maximal refinement of the proportions of varieties between themselves.

The heap which apparently offer a good effect of synthesis of the combustible burnt (broadening of the floristic spectrum) are difficult to date and the possibilities of interpretation in terms of economy of wood are limited. Indeed, these vast heaps of waste have they been formed in one year, or in a dozen or more ? A second problem is the "levelling" of the spectrum and the impossibility to detect the punctual variations or cycles which could reveal the modalities of management, the techniques and the evolution of the forest landscape. The variations between dominant taxa are diluted in the general mass of combustible used over a period difficult to evaluate.

A sampling on a larger portion of waste would be necessary along with those obtained in the corresponding exploitation zone.

## FIRST RESULTS AND INTERPRETATIONS

This preliminary study was realised on a total of 2429 charcoals spread over the type 1 and type 2 samples.

Despite the truncated character of the samples of type 1, data is integrated in the anthracological diagrams so as to complete provisionally certain chronological sequences comprised between the tenth and fourteenth centuries (FIG. 5). From the radiocarbon datings and the analysis of the operating dynamic, a relative chronology of the samplings is proposed.

The general diagram reveals the absolute dominance of taxa belonging to the arborean strata. *Larix-Picea* is very prominent in all the sequences, but *Pinus* type *P. sylvestris* overtakes it in the fourteenth century (end thirteenth-beginning fourteenth century?) in the portions of the "Combe Blanche" network and in a portion of the "Vieux Travaux" network (bottom of the mine). The frequencies of *Pinus* type *P. sylvestris* are weak in the workings exploited in the tenth to thirteenth century. There are all the same significant increases of its frequencies, notably in a portion of the "Lauzebrune" network ("salon") dated eleventh to twelfth century and in a portion of the "Saint Roch" network (schist shaft), dated end of thirteenth to beginning fourteenth century where they attain more than 20% (FIG. 5).

The predominance of *larix-Picea* is not tied to a technical choice as the heat value of the fires depended especially on the humidity of the wood and also their calibre, on the building up and tending of the fire and also the structure of the rock being mined. The anthracological analysis of the residual of firesetting in the antique mine of Hautech (Pyrenees) confirms this hypothesis. They present a large range of varieties used in firesetting: *Corylus avellana*, *Fagus sylvatica*, *Quercus* deciduous, *Prunus*, *Populus* (DUBOIS 1996: 39). The miners exploit in prority the varieties present in abundance in their local environnement, but this notion of proximity must be nuanced. It can mean the immediate neighbourhood or available woods near hand, but not necessarily situated in direct proximity to the mine. They could adapt the proprieties of the wood according to technical contraints such as varying the humidity level and also the size of logs (or faggots). Archaeological digs at the entry to the gallery of "Saint Roch" have revealed a work place set aside for the splitting and preparation of wood characterized by the numerous wood chips.

At the Fournel, the mines are situated between hilly shelves and the mountain, the growing area of *Pinus sylvestris*, moors with *Juniperus* and to a lesser degree *Quercus pubescens*. According to the diagrams, the forest exploited by the miners is composed essentially of *Larix-Picea*. The doubts associated with the reconnaissance of larch

| Periods | | Xth-XIth c. | | XIth-XIIth c. | | | | | | | | XIIIth-XIVth c. | | | XIVth c. | | | | Dating ? | | |
|---|---|---|---|---|---|---|---|---|---|---|---|---|---|---|---|---|---|---|---|---|---|
| Dating BP Intervalle corr. | | 893-1028 | | 1050-1150 | 909-1138 | | | | | | | 1206-1293 | | | 1286-1396 | | | | | | |
| Areas | | Lauzebrune Salon | Vieux Travaux Exhaure ancienne | Vieux Travaux Salle d'entrée A | Vieux Travaux Galerie aérage sud | Lauzebrune Erbstollen comblé | Saint Roch Coupe 2001 | Saint Roch | Saint Roch S1 : entrée sup. | Saint Roch S2 : entrée | | Saint Roch Chantier double | Combe Blanche | Vieux Travaux Puits des schistes | Combe Blanche | Vieux Travaux Ressaut 1001 | Combe Blanche | | Halde Blanche | Halde 2 | Halde Pinée |
| Type of sample NFI | | T.2 248 | T.1 100 | T.1 306 | T.1 25 | T.1 13 | T.2 150 | T.2 250 | T.1 485 | T.1 198 | | T.1 38 | T.2 100 | T.2 275 | T.2 100 | T.1 41 | T.2 100 | | T.2 638 | T.2. 108 | T.2 265 |

Taxa: Larix-Picea, Pinus type P. sylvestris, Abies sp., Quercus f.c., Salix sp., Alnus sp., Ulmus sp., Juniperus sp., Berberis sp., Prunus sp., Laburnum-cytisus, Fagus sylvatica, Indeter. Gymno.

*Inversion of dominant taxa*

*Broadering of floristic spectrum*

V. Py 2005

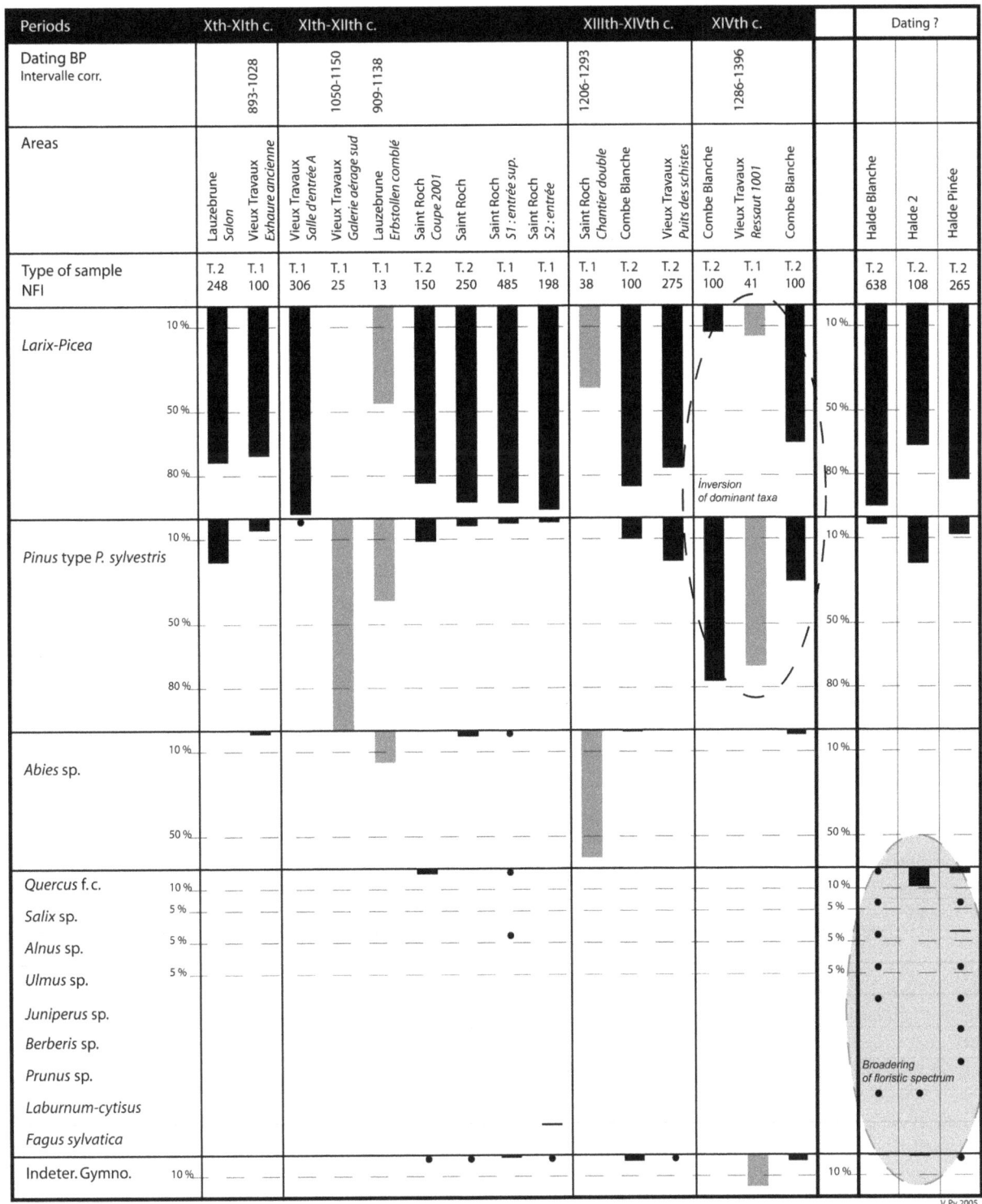

● < 1%  
Indeter. Gymno. = Indeterminable Gymnosperm  
NFI = Number of Fragments Identified  
■ Taxa relative frequencies (%).  
▨ Taxa relative frequencies (%) in "truncated" samples.

Indeterminable charcoals are not totalized for the calculation of relative frequencies.

**Figure 5.** General anthracological diagram of the Fournel mine.

(*Larix decidua*) and of the spruce (*Picea abies*), noted *Larix-Picea*, repose on the characteristics belonging to the anatomy of these two species, difficult to distinguish by means of compared anatomy. However, their respective ecology coupled with recognition of the diagnostic criteria of the larch, permits to lean towards *Larix* (TALON 1997).

Nowadays this variety is often in a pioneer phase and in false climax in the superior mountain shelves or in the subalpine inferior shelf in the locality of firs and spruces. The miners exploitation territory appears to be oriented towards the forests at the superior limit of the mountain level (to the *"ubac"*) and in the subalpine larch localities. This constatation is confirmed by the traditional practises of forest exploitation where it is preferred to clear the higher wood towards the bottom of the mountain. This provisioning can also correspond to the relative absence of combustible in the mountainous forests because *Pinus* type *P. sylvestris* is of generally feeble frequency. Its significant augmentation in the diagram of "Lauzebrune" can reflect a variation of the limits of the supply territory, reorientated punctually towards the mountain forests of the Southerly exposed versants.

The low frequencies of *Abies* projects the image of a pine locality in a colonisation phase, in the wooded massive of the *"ubac"*. The high frequencies registered at "Saint Roch" are related to an over representation of the taxa because the sample is localised and not representative. It should be at this time excluded from the paleoecological interpretation. The hypothesis of a degraded pine colony as early as the tenth century must be considered with much reserve. The feeble frequency of *Abies* could eventually correspond to the rigorous management of this species of first choice. In the Pyrenees, it has been subjected to a voluntary policy of conservation for workable wood: these are the *"bèdes"* or *"bedats"* of which the locality is totally independant of natural conditions (DAVASSE 2000). In the light of textual studies the conclusions are similar for the anthracological diagrams of the Catalonian Pyrenees (IZARD 1999). According to the regressive study of the high alpine texts, a policy of strict management of the pine localities has not yet been remarked. In the 18th century this variety was exploited in the same manner as pine and the larch for workable wood. This significant augmentation of the frequency of *Pinus* type *P. sylvestris* at the end of the 13th and beginning of the 14th century can reveal a localised variation in the time of supply strategies of wood for the miners. With the necessary reserves towards the interpretation of preliminary data, this modification of frequencies of dominant taxa can correspond to a new phase of exploitation of the mine after a period of non production, long enough to permit the regeneration of the pine forest, very dynamic. A second hypothesis, shows the problem of a very distinct variation of the area of exploitation which before was inclined towards high larch growths and reorientated towards forests of a dry mountainous type where *Pinus* type *P. sylvestris* could have been reserved at the expense of *Larix-Picea* higher in altitude.

*Quercus pubescens*, leader of the supramediterranean forest, does not appear, or at least very little in the anthracological diagrams (*Quercus deciduous*). This feeble representivity probably characterizes its reduced place in the ligneferous biomass available to the miners. The samples observed show narrow growth rings, characteristic of a wood which has developed in a xeric environnement (*"adret"*). With relative frequencies which reach the height of 2% at "Saint Roch" and of 9.3% in the "halde 2" sampling and in spite of limits tied to poor representivity of certain samples (type 1), the oak growth appears largely decapitated, at least since the 10th-11th centuries. It has given way to *Pinus* type *P. sylvestris*, a pioneering variety in the mountain and dry sub mountain forests. This interpretation implies a strong anthropisation of the versants at an early period. It may be the indirect reflection of a precocious apparition of *"blaches"* and oak plantations towards the 10th-11th centuries, perceptible in the written sources. This observation goes in the direction of the development of vinyards at this epoch (POGNEAUX 2001).

This early data permits a schematical outline of the types of forest management practised to obtain wood. As early as the 10th-11th centuries, the supply area opens in the high mountains forests knowing that the optimal localisation of *Larix* is actually situated between 1750 and 2100 m of altitude. This reflects probably a choice related to the forest availabilities and the accessibility of wooded areas for the laying of haul roads. *Pinus* type *P. sylvestris* appears to have been exploited more episodically with a significant peak at the end of the 13th beginning of 14th century. This variety develops generally lower in altitude. It is actually localised on the limestone massifs of the *"adret"* between 1200 and 1700 m of altitude *Pinus uncinata* takes up the relay (at the moment, growths of this pine are very localised). Its acquisition reflects an area which is punctually reorientated towards the *"adret"* massifs.

The diagrams show a wood collection which is not monospecific but adapted to the availabilities which may vary according to the state of wooded areas. The woodcutters could cut the trees in the areas (clearing or selective?) abandoned after partial or total deforestation to permit their regeneration. These same areas could be re-exploited several decades later knowing that about 50 years is needed for a larch situated in the superior limit of the forest to realize a diameter of 20 cm (PETITCOLAS *ET AL.* 1997). One can easily emit the hypothesis of management methods in a cyclic rotation form. Actually, no reglementation destined for wood management of the mining needs of the Middle Ages permits to validate a such hypothesis. Anyway, it is difficult to agree with the "traditional" vision of the medieval forest being attacked from all sides by a population unconcerned by its regeneration (SCLAFERT 1933).

## CONCLUSION

These first results show the pertinence of an anthracological approach in a mining context, validated notably by the ecological coherence of the diagrams. It permits to obtain new information on the methods concerning forest exploitation developed "upstream" of mining extraction. Above and beyond a strictly environmental aspect, it opens perspectives of research on the history of forest economy, valorisation of natural resources and operating systems of combustible material for alimentation of protoindustrial activities in the Middle Ages. From a paleoecological point of view, this brings to light the South-alpine medieval mountain forest still little known to environmentalists who until now tended to concentrate on the problem of altitudinal variations of the superior limit of the "dense" arborescent strata and the impact of the first agropastoral activities in high mountain regions (TALON 1996). Finally, it brings in some anchor points to study more finely the history of the larch and its expansion which is in close relation with the development of anthropical activities. Its management for the supply of firewood appears evident in the case of the Fournel mines.

This reflection constitutes a basis of theoric reconstruction of the evolution of an industrial landscape in the 10th-14th centuries. It implies looking from different angles at the paleoenvironmental approaches to varied species (palynology, dendochronology, analysis of lead) and on the archaeological data which characterises the structure and places.

_____

[1] The ancient silver mines of Fournel are, since 1992, the centre of an archaeological dig directed by Bruno Ancel and benefit concurrently from an aid programme of heritage development led by the town council of l'Argentière-la-Bessée.

[2] See contribution by Bruno Ancel and Vanessa Py in this book.

[3] The method of C. Marconnet is presented in the acts of the colloquium of Châteaudouble 2000, in press.

ACKNOWLEDGMENTS

The author wishes to thank particularly Mr Dave Aitken who undertook the fastiduous task of translation ; also Mr Ian Cowburn and Mr Andreas Hartmann-Virnich for the translation of the abstracts ; Mr Bruno Ancel for his scientific follow up of the archaeological operation concerning the study of firesetting backfill (granulometry and sedimentology).

Actually this study continues in the form of a doctorate thesis directed by Mr Michel Fixot. It will be part of a concerted action: *Burning knowledge, managment of combustible by the potters and miners of the meridional regions (11th-16th centuries)*, coordinated by Mrs Aline Durand and the Eclipse II programme coordinated by Mr Alain Veron: *Compared study of the evolution in high resolution of climatic events and anthropic activities in the French Southern Alps during the last two thousand years.*

REFERENCES

ANCEL B., 1998a.- La mine du Fournel (L'Argentière-La-Bessée, Hautes-Alpes, France): l'exploitation rationnelle aux Xᵉ-XIVᵉ siècles d'un filon de plomb argentifère, *in:* L. Brigo and M. Tizzoni (eds), *Actes du Congrès Européen Civezzano-Fornace*, 1995: 161-193.

ANCEL B., 1998b.- Techniques minières et maîtrise de l'espace dans les mines d'argent médiévales. Exemples de mines de plomb argentifère des Alpes du Sud (Xème-XIVème siècles), *in: Actes du Congrès d'Archéologie Médiévales (octobre 1996, Dijon)*, Paris, Editions Errance: 108-110.

ANCEL B., 2000.- Les anciennes mines des Hautes-Alpes (Ecrins, Queyras) et leur adaptation à l'environnement montagnard, *in:* G. Boetsch (ed.), *Les écosystèmes alpins, approches anthropologiques, Actes de l'université d'été 2000*, Gap, CDDP des Hautes Alpes: 88-95.

ANCEL B., IN PRESS.- La mine d'argent du Fournel à L'Argentière-la-Bessée (Hautes-Alpes): méthodologie et bilan 1991-2001, *in: Mine et métallurgie en Provence et dans les Alpes du Sud de la Préhistoire au XXe siècle: reconversion industrielle et enjeux culturels, actes du colloque Châteaudouble* (2001).

ANCEL B., MARCONNET C., KAMMENTHALER E., IN PRESS (a).- La mine d'argent du Fournel à l'Argentière-la-Bessée: bilan des fouilles programmées 1992-2001, *in: Actes du colloque de St Clément-les-Places.*

ANCEL B., PY V., MARCONNET C., IN PRESS (b).- De l'usage minier du feu: à l'interface homme et environnement. Sources et expérimentations, *Cahier d'Histoire des Techniques*, Publications de l'Université de Provence.

BADAL GARCIA E., 1990.- Méthode de prélèvement et paléoécologie du Néolithique d'après les charbons de bois de «la Cova de les Cendres» (Alicante, Espagne), *in:* T. Hackens, A.V. Munaut, and C. Till (eds), *Wood and Archaeology (Bois et Archéologie). First European Conference (Louvain-la-Neuve, October 2nd-3rd 1987)*, Strasbourg, PACT, 22: 231-243.

BERG B.I., 1992.- Les techniques d'abattage à Kongsberg (Norvège) du XVIIe au XIXe siècle: pointerolle, travail au feu et tir à la poudre, *in: Les techniques minières de l'Antiquité au XVIIIe siècle, Actes du colloque international sur les ressources minières et l'histoire de leur exploitation de l'Antiquité à la fin du XVIIIe siècle (Strasbourg, avril*

*1988)*, Paris, Editions du C.T.H.S.: 55-76.

CHABAL L., 1982.- *Méthodes de prélèvement des bois carbonisés protohistoriques pour l'étude des relations homme-végétation*, DEA, Université de Montpellier II, 54 p.

CHABAL L., 1988.- Pourquoi et comment prélever les charbons de bois pour la période antique: les méthodes utilisées sur le site de Lattes (Hérault), *Lattara*, 1: 187-222.

CHABAL L., 1991.- *L'homme et l'évolution de la végétation méditerranéenne, des âges des métaux à la période romaine: recherches anthracologiques théoriques, appliquées principalement à des sites du bas Languedoc*, Doctorat, Université de Montpellier II, 435 p.

CHABAL L., 1992.- La représentativité paléo-écologique des charbons de bois archéologiques issus du bois de feu, *in*: J.-L. Vernet (ed.), *Les Charbons de bois, les anciens écosystèmes et le rôle de l'homme. Colloque organisé à Montpellier en décembre 1991*, Paris, Editions Bulletin de la Société Botanique Française: 213-236 (Actualités Botaniques. 1992-2/3/4).

CHABAL L., 1997.- *Forêts et sociétés en Languedoc (Néolithique final, Antiquité tardive). L'anthracologie, méthode et paléoécologie*, Paris, Editions de la Maison des Sciences de l'Homme, 189 p. (Document d'Archéologie Française, 63).

DAVASSE B., 2000.- *Forêts, charbonniers et paysans dans les Pyrénées de l'Est, du Moyen Age à nos jours. Une approche géographique de l'histoire de l'environnement*, Toulouse, GEODE, 287 p.

DUBOIS C., 1996.- L'ouverture par le feu dans les mines: histoire, archéologie et expérimentation, *Revue d'Archéométrie*, 20, 1996: 33-46.

HEINZ C., 1988.- *Dynamique des végétations holocènes en Méditerranée Nord-Occidentale d'après l'anthracoanalyse de sites préhistoriques: méthodologie et paléoécologie*, Doctorat, Université de Montpellier II, 275 p.

IZARD V., 1999.- *Les montagnes du fer. Eco Histoire de la métallurgie et des forêts dans les Pyrénées méditerranéennes (de l'Antiquité à nos jours). Pour une histoire de l'environnement*, Doctorat, Université de Toulouse II, 752 p.

MARCONNET C., 1994.- *La préparation mécanique du minerai de galène, au XIX*e *siècle, sur le site du Fournel, à partir d'une étude sédimentologique des restes de traitement.* DEA, Université de Paris I, 64 p.

MEYER D., 1981.- *La végétation des vallées de Vallouise, du Fournel et de la Biaysse (Pelvoux Oriental, Hautes-Alpes). Analyse phytosociologique et phytogéographique des étages collinéen, montagnard et subalpin*, Doctorat, Université d'Aix-Marseille I, 176 p.

PETICOLAS V., ROLLAND C., MICHALET R., 1997.- Croissance de l'épicéa, du mélèze, du Pin cembro et du Pin à crochets en limite supérieure de la forêt dans quatre régions des Alpes françaises, *Annales des Sciences Forestières*, 54 (8): 731-745.

POGNEAUX N., 2001.- *Le vignoble d'altitude Bacchus y trouva un royaume!*, L'Argentière-la-Bessée, Editions du Fournel, 96 p.

RAMEAU J.-C., MANSION D., DUMÉ G., 1993.- *Flore forestière française: guide écologique illustré. 2- Montagnes*, Paris, Institut pour le développement forestier/Ministère de l'agriculture et de la pêche/Direction de l'espace rural et de la forêt/Ecole nationale du génie rural des eaux et forêts, 2421 p.

SCLAFERT T., 1933.- A propos du déboisement des Alpes du Sud, *Annales de Géographie*, XLII: 266-277 and 350-360.

TALON B., 1996.- *Evolution des zones supra-forestières des Alpes du Sud-Occidentales françaises au cours de l'holocène: analyse pédoanthracologique*, Doctorat, Université d'Aix-Marseille III, 186 p.

TALON B., 1997.- Etude anatomique et comparative de charbons de bois de *Larix decidua* Mill. et de *Picea abies*, *Science de la vie*, 320: 581-588.

# CHARCOAL ANATOMY POTENTIAL, WOOD DIAMETER AND RADIAL GROWTH

Alexa DUFRAISSE

Laboratoire de Chrono-écologie, UMR 6565 CNRS,
Université de Franche-Comté
16 route de Gray F-25030 Besançon cedex (France)
alexa.dufraisse@univ-fcomte.fr

**ABSTRACT**: The study of charcoal fragments from archaeological contexts is often limited to the identification of wood taxa. However, the dendrological examination of charcoal provides much information about physiological and health state of wood, wood diameter, wood growth, which offer valuable data on palaeoecology and on firewood management. Therefore, this paper proposes a brief review of the charcoal anatomy potential. It also proposes two approaches; the first one allows a reconstruction of the burnt wood diameter used in the past which correspond to the relative distribution of the diameter classes in an archaeological sample. The second method we present is an eco-anatomical approach based on the measure of the tree-ring width adapted to fragmented charcoal corpus.
**KEY WORDS**: Wood Charcoal, Dendrology, Eco-anatomy, Wood Diameter, Radial Growth

**RÉSUMÉ**: L'étude des charbons de bois provenant de contexte archéologique est souvent limitée à l'identification des essences. Cependant, l'examen dendrologique peut apporter de précieuses informations sur l'état physiologique et phénologique du bois par exemple, les diamètres utilisés, les conditions de croissance, autant d'informations indispensables à l'interprétation des spectres en terme de paléoécologie et de gestion du bois de feu. Par conséquent, nous proposons tout d'abord un bref rappel du potentiel anatomique des charbons de bois. Puis, deux nouvelles approches seront présentées, la première permet une reconstruction des diamètres de bois brûlés à partir de la distribution relative des classes de diamètres représentées dans un échantillon. La seconde méthode est une approche éco-anatomique fondée sur l'étude des largeurs adaptée aux corpus fragmentés de charbons de bois.
**MOTS-CLÉS**: Charbons de bois, Dendrologie, Eco-anatomie, diamètre de bois, croissance radiale

**ZUSAMMENFASSUNG**: Die Untersuchung der Holzkohlenfragmente aus archäologischem Kontext beschränkt sich oft auf die Identifikation der Taxa. Die dendrologische Untersuchung der Holzkohlen kann jedoch darüber hinaus wertvolle Informationen über den physiologischen und phänologischen Zustand des Waldes liefern, z.B. über die benutzten Holzdurchmesser und über die Wachstumsbedingungen. Dadurch erhält man paläoökologisch und feuerholztechnisch unverzichtbare Informationen zur Interpretation des Artenspektrums. Das vorliegende paper bietet zuerst einen kurzen Abriss der anatomischen Möglichkeiten von Holzkohlen. Danach werden zwei neue Ansätze präsentiert. Der erste erlaubt die Rekonstruktion der Durchmesser von verbranntem Holz, ausgehend von der relativen Verteilung der Durchmesserklassen in einer kleinen Stichprobe. Der zweite Ansatz ist eine ökoanatomische Annäherung, bei der die Jahrringbreiten mit den Holzdurchmessern kombiniert werden.
**STICHWORTE**: Holzkohlen, Dendrologie, Öko-Anatomie, Holzdurchmesser, radiales Wachstum

Over the last ten years, new analytical tools have been developed in the anthracological field so as to go beyond the classical palaeoecological pattern of interpretation (see the proceedings of the second international meeting of anthracology, THIÉBAULT 2002). New approaches are being developed, especially focused on the dendrological examination of charcoal. Indeed, dendrological examination is a primordial step in charcoal analysis. It allows discovery of much information on the human use of wood, firewood gathering modes, and the human impact of exploitation on woodland. A vast methodological study has been undertaken on both fundamental anatomy and on eco-anatomical features of wood, methods which are still insufficiently developed for the study of charcoal from archaeological contexts for which the classical methods usually used in dendrology cannot be directly applied.

After a brief review of the potential of charcoal anatomy analysis, this paper will focus on both themes that we have developed since our PhD thesis (DUFRAISSE 2002): wood diameter and radial growth.

## CHARCOAL ANATOMY POTENTIAL

The dendrological examination of charcoal can provide much information about firewood management, such as the species used, the physiological state of wood (dry or green) or the state of health of the wood (healthy or weathered).

### Species, combustion proprieties and wood selection

The identification of the wood taxa of charcoal fragments, most often to the species level, is made through observation of the three natural planes of wood by means of binoculars and an incident-light microscope, comparisons with a wood anatomy atlas (GREGUSS 1959, JACQUIOT 1955, JACQUIOT ET AL. 1973, SCHWEINGRUBER 1990a, b...), and comparative collections. However, precise identification of charred wood is sometimes problematic. Indeed, key features such as size and dimensions of anatomical structures are deformed due to carbonisation processes (shrinkage, craking, etc.).

These identifications allow establishment of a floristic list which forms a basis for interpretation. When the number of taxa is small, as is often the case in craft or industrial contexts, it is easy to propose hypotheses as to firewood management. Nevertheless, floristic lists most often contain between 20 and 30 taxa, which is a considerable number for a European forest. In this case, charcoal interpretation is based on the "Principle of the Least Effort" (SHAKLETON AND PRINS 1992). This means that firewood collection took place in areas close to settlements and that all species were collected in accordance to their proportion in the environment. Consequently, charcoal assemblage allows a reconstruction of forest composition and changes over time.

However, firewood collection and management is also a reflection of the social organisation of a settlement (DUFRAISSE 2005, DUFRAISSE ET AL. IN PRESS). Thus, charred wood analysis constitutes a good indicator of economic behaviour as seeds remain. It is therefore possible to study the co-evolutions of social, technical, economic and environmental contexts through charcoal analysis. In addition, firewood collection can be directed by other selection criteria as is shown by ethnology (SMART AND HOFFMAN 1987); firewood selection is also based on parameters such as the physiological and weathered state of wood, or its diameter. Anatomical deformations

of charred wood constitute sensitive tools for the determination of firewood collection customs in a group (THÉRY-PARISOT 2001).

## Green wood or dry wood

Two main anatomical features provide information about the physiological state of wood just before carbonisation.

The first is the local presence of a cellular collapse zone, observable on the transversal plane and which may be formed during the carbonisation of green wood (FIG. 1a) (THÉRY-PARISOT 2001). The presence of such features is generally, and rightly, interpreted as the collection (and the use) of green wood. However, these features can also appear during the life of a tree, due, for example, to water stress (frost or dry year), which makes diagnosis more difficult. However, in the last case, the collapse zones are concentrated in one tree ring and not distributed all along the transversal plane.

Concerning interpretation, on the one hand, experimentation by I. Théry-Parisot (1998) underlines that 100% of green wood gives only between 15% and 20% of charcoal fragments with a collapse zone. On the other hand, the absence of such features is not sufficient to refute the use of green wood.

The second anatomical feature, and one perhaps better known, is the radial cracks which also occur during carbonisation of green wood (FIG. 1b) (MARGUERIE 1992). Their occurrence varies from one species to another. As for the collapse zone, these features can also be formed by a mechanical phenomenon before carbonisation. However, it can be observed that in a sample, the proportions of the collapse zone increase with the proportions of the radial cracks (DUFRAISSE 2002). Consequently, we think that a combination of both features can provide a good estimation of the physiological state of the wood (FIG. 1c).

## Healthy wood or weathered wood

Dead wood can be of physiological, parasitical or climatic origin. The difference between dead and green wood is the proportion of free water. However, dead wood may also be weathered but not necessarily.

The use of weathered wood can be determined by the presence of fungi in the vessels of the wood. Indeed, mycelium cannot invest a wood charcoal composed of 100% of carbon. Thus these fungi invested the tree when it was alive. In addition, a few experiments show that fungi are perfectly preserved after carbonisation of the wood (THÉRY-PARISOT 1998). The wood structure may also be deformed; for example, the compression of a weathered wood such as Fagus leads to a transversal

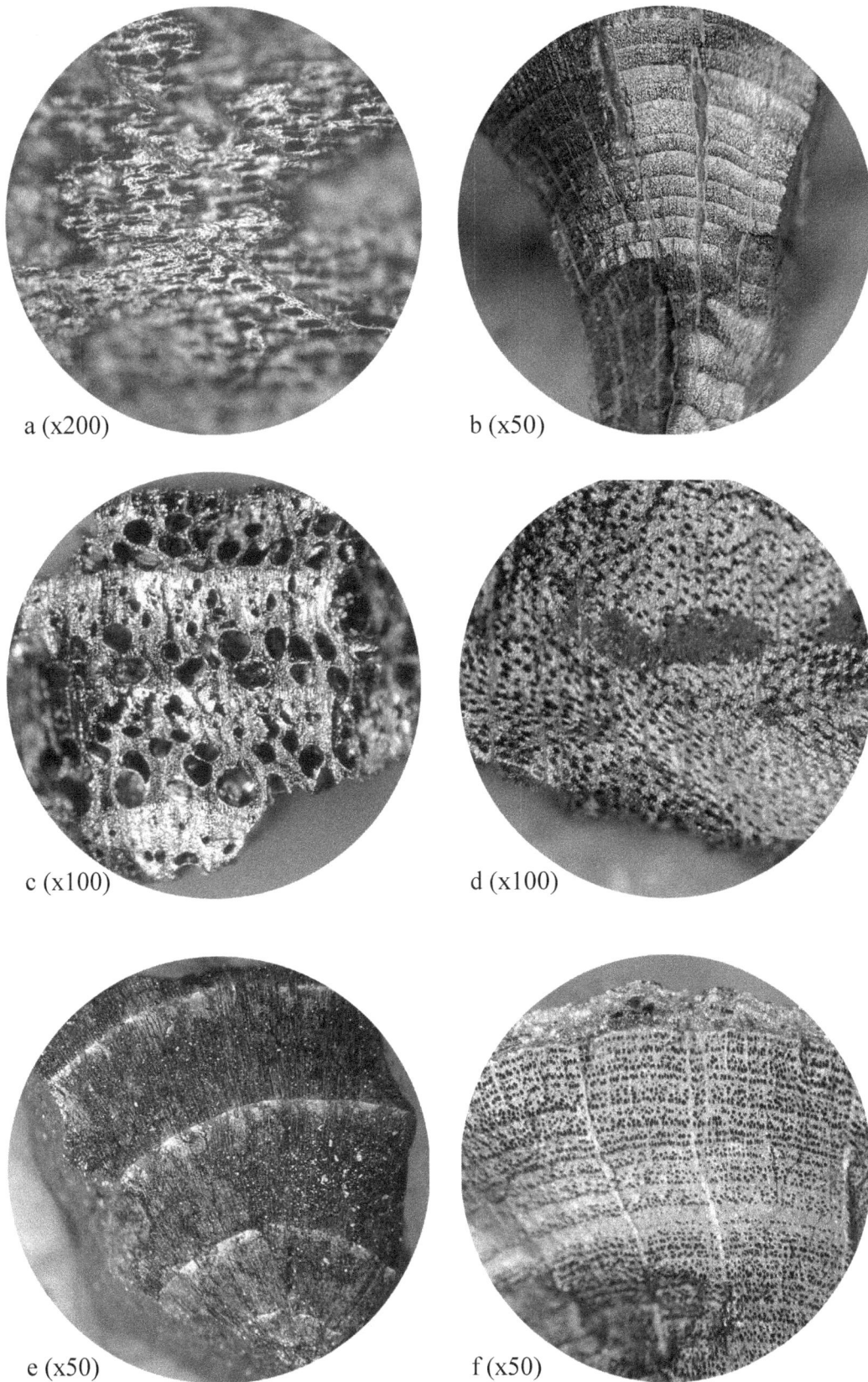

**Figure 1.** Anatomical features on archaeological wood charcoal reflecting human uses. a- cellular collapse zone on *Fagus* (x200); b- radial cracks on a branch of *Fagus* (x50); c- combination of cellular collapse zone and radial cracks on *Fraxinus* (x100); d- scares tissues due to Aphids on Pomoideae (x100); e- charcoal fragment from a trunk, *Abies* (x50); f- charcoal fragment from a twig of *Fagus* (x50).

plane with sinuous rays. Concerning the interpretation, the presence of mycelium indicates the use of dead wood and not the collection of dead wood; indeed fungi can invest a wood during a period of drying after its acquisition.

Presence of thyloses in some deciduous trees may indicate the difference between heartwood and sapwood. However, this feature is also a self-defence mechanism of the tree to protect itself from a fungi attack. Lastly, wood structure can also show traces of xylophagous insects such as gallery holes and scar tissue (FIG. 1d).

## Trunk, twigs, roots...

Through dendrological examination of wood charcoal, it is possible to observe anatomical structures and thus to determine from which part of the tree the firewood comes (SCHWEINGRUBER 1990b). For example, in a root, there is no pith and the size of the cell is clearly larger than in a trunk. In contrast, in a branch, there is pith and the size of the cells is clearly smaller than in a trunk (FIG. 1e and f). Finally, these types of data provide valuable information on the environment, human activities and wood use: wood collection focussed on branches can modify the architectural form of the trees (see THIÉBAULT, BERNARD *ET AL.* in this book) which will in turn structure the spatial and the temporal organisation of a wood stand. Better yet, the presence of roots in charcoal samples from Neolithic sites can be an additional proof in demonstrating the activity of "rooting-out" linked to the clearing of new fields.

## WOOD DIAMETER

### Laboratory method

Wood diameter is a criterion of firewood selection as important as the species or the physiological state of the wood. According to the protocol established by other authors (such as MARGUERIE 1992, LUDEMANN AND NELLE 2002, MARCONETTO 2002, NELLE 2002...), wood diameters can be determined by the measure of the tree ring bending using a diameter stencil (a test card) and a magnifying glass (or a binocular lens).

For our study, based on more than 20,000 charcoal fragments from Neolithic sites (DUFRAISSE 2002 and 2005), only charcoal pieces larger than 4 mm were taken into account for identification (rare anatomical characteristics may be absent from specimens smaller than 4 mm). However, we wondered if the diameter class depends on the size of the charcoal fragment. We thus undertook a statistical test of correlation (Spearman's Rank Correlation Test) which indicated a very weak correlation between the size of the fragment and the measured diameter class ($r = 0.35$ to $r = 0.49$ following the samples) (DUFRAISSE 2002).

### Rank data and interpretation

According to the relative vagueness of the measure (several values may correspond to one charcoal fragment) and taking into account shrinking during carbonisation (about 15 - 20% along the radial axis but assumed to be

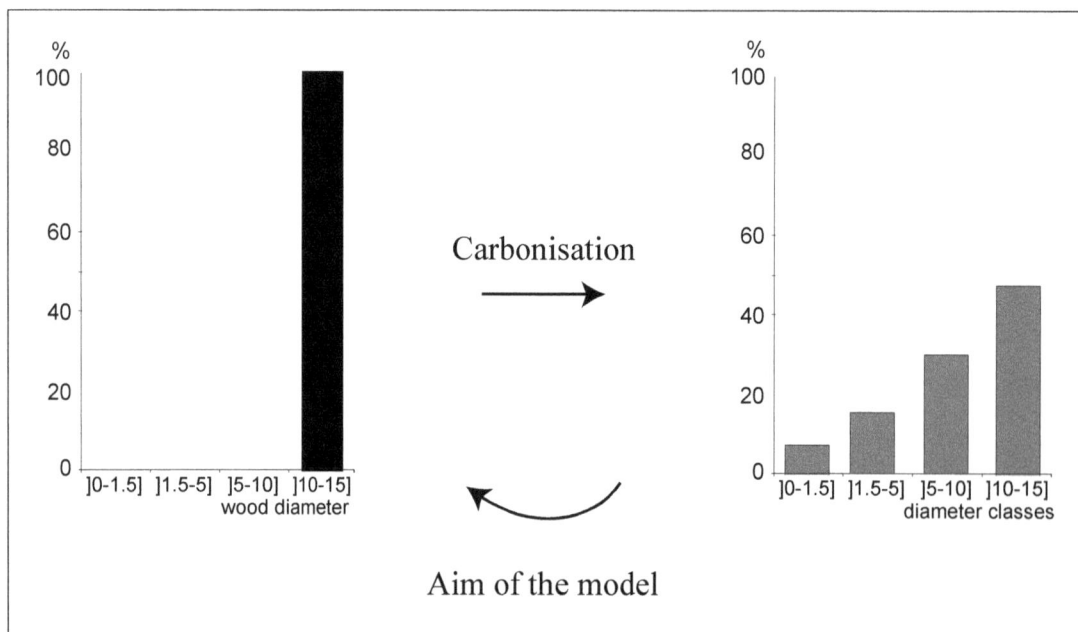

**Figure 2.** Theoretical distribution of the diameter classes after the carbonisation of a fire-log of 15 cm in diameter.

proportional), we chose to put the data into equal classes except for the first which is easier to determine: ]0-1.5], ]1.5-5], ]5-10], ]10-15], ]15-20] and ]20-25]. Each taxa was considered independently.

The measure of tree-ring bending can be interpreted as follows: if we burn a fire-log of 15 cm diameter, many charcoal fragments result whose tree ring bending measurement lies between 1 and 15 cm in diameter. Thus, the tree ring curvature indicates that the analysed fragment is situated within a specific zone in the trunk. The data do not indicate the percentages of each diameter used in the past, but rather the distribution of the wood diameter classes in the archaeological sample corresponding to a special (and unknown) arrangement of diameters

(Fig. 2). In this way, one can compare the results between different sites. However, because we wanted to know wood diameters used in the past, we developed a model based on mathematics and geometry. It allows estimation of theoretical diameter proportions before carbonisation (or used in the past) which correspond to the proportions of diameter classes after carbonisation (or in the archaeological sample).

## Elaboration of a model

### Basic considerations

For the elaboration of this model, we considered that a

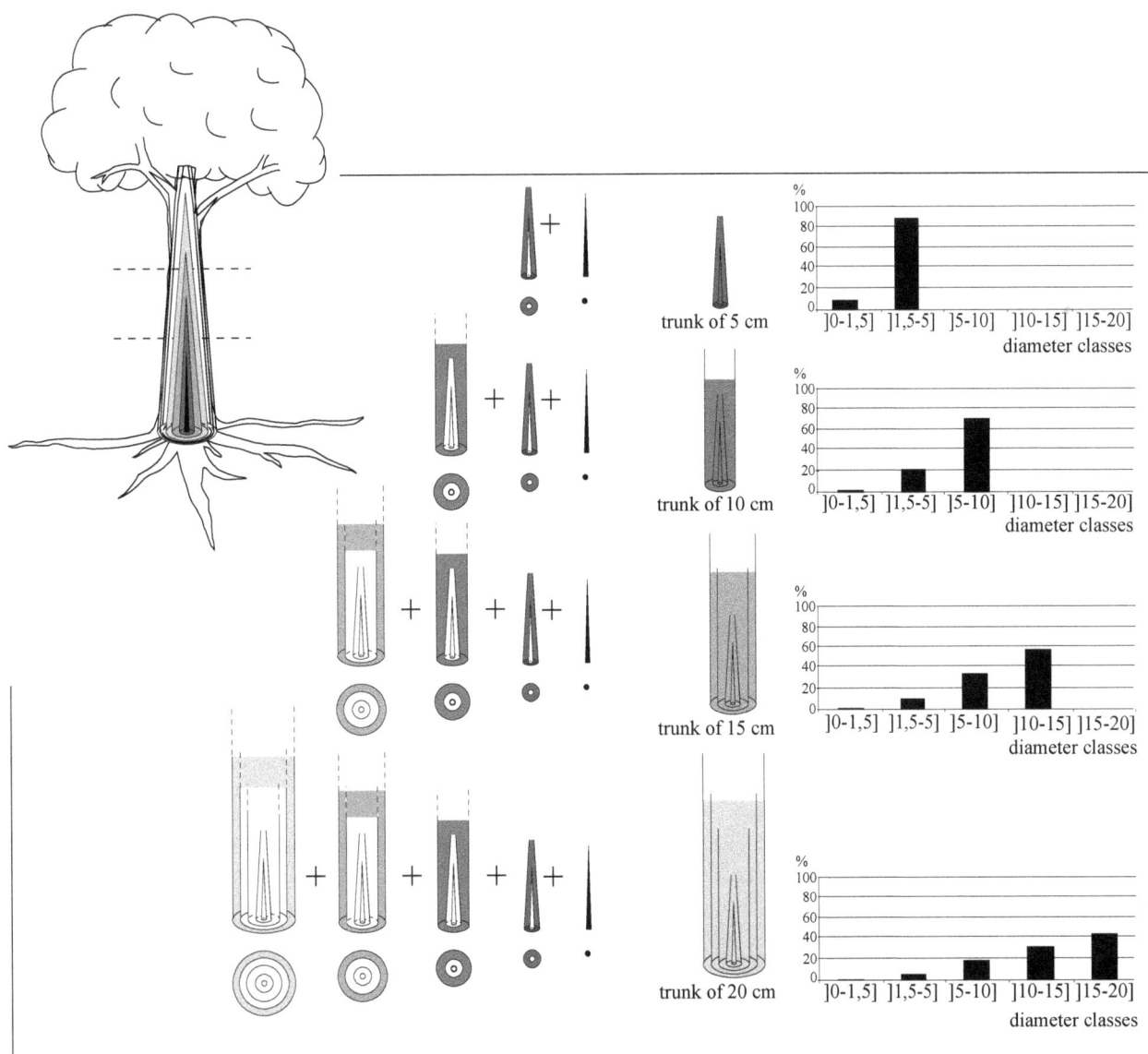

**Figure 3.** Principle of the model (a trunk is formed by a stacking of empty cylinders whose thickness is defined by the diameter classes) and volume distribution of the hollow cylinders for each wood diameter (or trunk).

| Diameter classes (cm) | ]0-1.5] | ]1.5-5] | ]5-10] | ]10-15] | ]15-20] | ]20-25] |
|---|---|---|---|---|---|---|
| Fire-log diameter (cm) | 1.5 | 5 | 10 | 15 | 20 | 25 |
| Ray (cm) | 0.75 | 2.5 | 5 | 7.5 | 10 | 12.5 |
| Volume V= r²x(3,14h) (cm³) | 0.5625 | 6.25 | 25 | 56.25 | 100 | 156.25 |
| Volume V' (hollow cylinder) | 0.56 | 5.69 | 18.75 | 31.25 | 43.75 | 56.25 |

**Figure 4.** Calculations of the volume of each hollow cylinder according to the defined diameter classes: ]0-1.5], ]1.5-5], ]5-10], ]10-15], ]15-20], and ]20-25].

| | | Volume distribution of hollow cylinders (%) | | | | | |
|---|---|---|---|---|---|---|---|
| | | ]0-1.5] | ]1.5-5] | ]5-10] | ]10-15] | ]15-20] | ]20-25] |
| Fire-log diameter | 1.5 cm | 100.0 | | | | | |
| | 5 cm | 9.0 | 91.0 | | | | |
| | 10 cm | 2.3 | 22.8 | 75.0 | | | |
| | 15 cm | 1.0 | 10.1 | 33.3 | 55.6 | | |
| | 20 cm | 0.6 | 5.7 | 18.8 | 31.3 | 43.8 | |
| | 25 cm | 0.0 | 3.6 | 12.0 | 20.0 | 28.0 | 36.0 |

**Figure 5.** Distribution of the relative volume of the hollow cylinders according to the diameter of the trunk (here from 1,5 cm to 25 cm in diameter).

trunk is formed by a stacking of "empty cylinders" (or cones)1 with a thickness that corresponds to the previous diameter classes. This model is based on the relative volume of each hollow cylinder that composed a trunk (or twigs, branches...) with a diameter of 5 cm, 10 cm, 15 cm, 20 cm, etc. (FIG. 3). Consequently, the volume of all charcoal fragments was systematically measured. In addition, as the model is based on relative volume, all data are explained in percentages.

Thus, we first calculated the relative volume of each hollow cylinder according to the diameter classes (FIG. 4). For this, one needs:
- the diameter of the fire log;
- the ray;
- the volume of the full cylinder (V); for this step, we considered that each cylinder which forms the fire-log has the same length;
- the volume of the hollow cylinder (V').

These values were then used to calculate the volume distribution of the hollow cylinders that compose a fire-log whose diameter is identified (FIG. 5). It can be observed that, for the diameter classes we have chosen, the hollow cylinder that corresponds to the greatest diameter class has the highest percentages (which is not the case for uneven diameter classes).

In order to re-use these distributions to estimate the theoretical distribution of wood diameters corresponding to the distribution of diameter classes in the archaeological sample, we established a "model table" (for the full

version, see DUFRAISSE 2002) where the relative volume of a hollow cylinder is explained according to the theoretical percentage of wood diameter (FIG. 6). For example, the theoretical distribution of diameter classes for a sample composed 100% of fire-logs of 15 cm diameter before carbonisation will be: 1% for ]0-1.5], 10.1% for ]1.5-5], 33.6 for ]5-10] and 55.6% for ]5-10]. But if this size of fire-log before carbonisation composes only 13% of the sample, the theoretical distribution of the diameter classes remains proportional but with lower relative values: 0.1% for ]0-1.5], 1.3% for ]1.5-5], 4.3% for ]5-10] and 7.7% for ]5-10].

*Example*

Practically, after charcoal analysis, we have only the relative distribution of the diameter classes calculated with the volume of the charcoal fragment in the archaeological sample and we look for the proportions of wood diameter before the carbonisation.

As an example of application of this model, we chose the results of tree ring bending of *Fraxinus* at Chalain 19. The results are explained with the number and the volume of charcoal fragments (FIG. 7).
To use the model, we must begin with the greatest diameter class represented. Here it is the class 15-20 cm with a percentage of about 32%.
We then look in the model table for the value of 32% for a wood diameter of 20 cm in the class of 15-20 cm, and we report the values of the theoretical distribution of the different diameter classes.

| | | 5 cm | | 10 cm | | | 15 cm | | | | 20 cm | | | | |
|---|---|---|---|---|---|---|---|---|---|---|---|---|---|---|---|
| | | DRV hc | | DRV hc | | | DRV hc | | | | DRV hc | | | | |
| | | ]0-1.5] | ]1.5-5] | ]0-1.5] | ]1.5-5] | ]5-10] | ]0-1.5] | ]1.5-5] | ]5-10] | ]10-15] | ]0-1.5] | ]1.5-5] | ]5-10] | ]10-15] | ]15-20] |
| | 1 | 0,1 | 0,9 | 0,0 | 0,2 | 0,8 | 0,0 | 0,1 | 0,3 | 0,6 | 0,0 | 0,1 | 0,2 | 0,3 | 0,4 |
| | 2 | 0,2 | 1,8 | 0,0 | 0,5 | 1,5 | 0,0 | 0,2 | 0,7 | 1,1 | 0,0 | 0,1 | 0,4 | 0,6 | 0,9 |
| | 3 | 0,3 | 2,7 | 0,1 | 0,7 | 2,3 | 0,0 | 0,3 | 1,0 | 1,7 | 0,0 | 0,2 | 0,6 | 0,9 | 1,3 |
| | 4 | 0,4 | 3,6 | 0,1 | 0,9 | 3,0 | 0,0 | 0,4 | 1,3 | 2,2 | 0,0 | 0,2 | 0,8 | 1,2 | 1,8 |
| | 5 | 0,5 | 4,6 | 0,1 | 1,2 | 3,8 | 0,1 | 0,5 | 1,7 | 2,8 | 0,1 | 0,3 | 1,0 | 1,6 | 2,2 |
| | 6 | 0,5 | 5,5 | 0,1 | 1,4 | 4,5 | 0,1 | 0,6 | 2,0 | 3,4 | 0,1 | 0,4 | 1,1 | 1,9 | 2,6 |
| | 7 | 0,6 | 6,4 | 0,1 | 1,6 | 5,3 | 0,1 | 0,7 | 2,3 | 3,9 | 0,1 | 0,4 | 1,3 | 2,2 | 3,1 |
| | 8 | 0,7 | 7,3 | 0,2 | 1,8 | 6,0 | 0,1 | 0,8 | 2,6 | 4,5 | 0,1 | 0,5 | 1,5 | 2,5 | 3,5 |
| | 9 | 0,8 | 8,2 | 0,2 | 2,1 | 6,8 | 0,1 | 0,9 | 3,0 | 5,0 | 0,1 | 0,5 | 1,7 | 2,8 | 4,0 |
| | 10 | 0,9 | 9,1 | 0,2 | 2,3 | 7,5 | 0,1 | 1,0 | 3,3 | 5,6 | 0,1 | 0,6 | 1,9 | 3,1 | 4,4 |
| | 11 | 1,0 | 10,0 | 0,2 | 2,5 | 8,3 | 0,1 | 1,1 | 3,6 | 6,2 | 0,1 | 0,7 | 2,1 | 3,4 | 4,8 |
| | 12 | 1,1 | 10,9 | 0,2 | 2,8 | 9,0 | 0,1 | 1,2 | 4,0 | 6,7 | 0,1 | 0,7 | 2,3 | 3,7 | 5,3 |
| | 13 | 1,2 | 11,8 | 0,3 | 3,0 | 9,8 | 0,1 | 1,3 | 4,3 | 7,3 | 0,1 | 0,8 | 2,5 | 4,0 | 5,7 |
| | 14 | 1,3 | 12,7 | 0,3 | 3,2 | 10,5 | 0,1 | 1,4 | 4,6 | 7,8 | 0,1 | 0,8 | 2,7 | 4,3 | 6,2 |
| | 15 | 1,4 | 13,7 | 0,3 | 3,5 | 11,3 | 0,2 | 1,5 | 5,0 | 8,4 | 0,2 | 0,9 | 2,9 | 4,7 | 6,6 |
| | 16 | 1,4 | 14,6 | 0,3 | 3,7 | 12,0 | 0,2 | 1,6 | 5,3 | 9,0 | 0,2 | 1,0 | 3,0 | 5,0 | 7,0 |
| | 17 | 1,5 | 15,5 | 0,3 | 3,9 | 12,8 | 0,2 | 1,7 | 5,6 | 9,5 | 0,2 | 1,0 | 3,2 | 5,3 | 7,5 |
| | 18 | 1,6 | 16,4 | 0,4 | 4,1 | 13,5 | 0,2 | 1,8 | 5,9 | 10,1 | 0,2 | 1,1 | 3,4 | 5,6 | 7,9 |
| | 19 | 1,7 | 17,3 | 0,4 | 4,4 | 14,3 | 0,2 | 1,9 | 6,3 | 10,6 | 0,2 | 1,1 | 3,6 | 5,9 | 8,4 |
| | 20 | 1,8 | 18,2 | 0,4 | 4,6 | 15,0 | 0,2 | 2,0 | 6,6 | 11,2 | 0,2 | 1,2 | 3,8 | 6,2 | 8,8 |
| | 21 | 1,9 | 19,1 | 0,4 | 4,8 | 15,8 | 0,2 | 2,1 | 6,9 | 11,8 | 0,2 | 1,3 | 4,0 | 6,5 | 9,2 |
| | 22 | 2,0 | 20,0 | 0,4 | 5,1 | 16,5 | 0,2 | 2,2 | 7,3 | 12,3 | 0,2 | 1,3 | 4,2 | 6,8 | 9,7 |
| | ... | ... | ... | ... | ... | ... | ... | ... | ... | ... | ... | ... | ... | ... | ... |
| | ... | ... | ... | ... | ... | ... | ... | ... | ... | ... | ... | ... | ... | ... | ... |
| | ... | ... | ... | ... | ... | ... | ... | ... | ... | ... | ... | ... | ... | ... | ... |
| | 100 | 9.0 | 91.0 | 2.3 | 22.8 | 75.0 | 1.0 | 10.1 | 33.3 | 55.6 | 0.6 | 5.7 | 18.8 | 31.3 | 43.8 |

*(Left column label, rotated: Theoretical percentage of wood diameter before the carbonisation. Top spanning header: Fire-log diameters)*

DRV hc : Distribution of the Relative Volume of the hollow cylinders (according to the diameter classes)

**Figure 6.** The model table. This table shows the relative distribution of the hollow cylinder according to the trunk diameter and the diameter classes which compose it. These relative distributions are expressed according to the theoretical percentage of the wood diameter before the carbonisation (left colon).

We consider now the following diameter class (10-15 cm) of 13.4%. But, with such a representation of the greatest class in the archaeological sample, we should have a theoretical value of the empty cylinder of around 22.3%, but we only have 13.4%. This class is therefore under-represented which is probably due to post-depositional conditions and we consider that no fire-log of 15 cm diameter was used.

The next diameter class, from 5 to 10 cm is represented with 41.3%. Among this percentage, a portion belongs to the biggest size of wood diameter (13.4%). Thus, we have to deduce this previous value (27% left) before using the model table. In the latter, we must seek the value of 27% for a wood diameter of 10 cm in the class of 5-10 cm and we report the theoretical distribution of the diameter classes.

We consider now the diameter class of 1.5-5 cm, represented by 13% of the archaeological sample. But, a portion of this value belongs to the preceding class and to the largest one. As above, we deduce them from 13%, which leaves 0.8%. We seek this value for a wood diameter of 5 cm in the class of 1.5-5 cm and we report the theoretical distribution of the diameter classes.

Lastly, for the smallest diameter class, 0-1.5 cm, according to the representation of the preceding classes, we should have at least 1.5% in the archaeological sample but it is not represented, probably due to the post-depositional conditions.

Consequently, with this model, we can, on the one hand, offer an estimation of the different sizes of wood diameter used in the past, and, on the other hand, check the representation of the diameter classes in the archaeological sample which depends partially on the post-depositional conditions.

Due to this bias, the total percentage of the diameter classes and the theoretical percentages of wood diameter used in the past are sometimes more or less than 100% and must be corrected (FIG 8).

Figure 9a allows comparison of the distribution of the diameter classes in the archaeological sample, corrected by the model. It can be observed that some classes are under or over-represented. For example, we can note the absence of the smallest diameter class in

| Diameter classes (cm) | | | | | | |
|---|---|---|---|---|---|---|
| | ]0-1.5] | ]1.5-5] | ]5-10] | ]10-15] | ]15-20] | total |
| fragments | 0,0 | 11,0 | 9,0 | 10,0 | 4,0 | 34,0 |
| % | 0,0 | 32,4 | 26,5 | 29,4 | 11,8 | 100% |
| Volume (cm³) | 0,0 | 7,6 | 23,2 | 7,5 | 17,8 | 56,0 |
| % | *0,0* | *13,6* | *41,3* | *13,4* | *31,7* | 100% |

fire-log of 20 cm diameter

| | ]0-1.5] | ]1.5-5] | ]5-10] | ]10-15] | ]15-20] |
|---|---|---|---|---|---|
| 72% | 0,7 | 4,3 | 13,7 | 22,3 | **31,7** |

*13,4*
8.9% lacking

fire-log of 15 cm

| | ]0-1.5] | ]1.5-5] | ]5-10] | ]10-15] |
|---|---|---|---|---|
| 0% | 0 | 0 | 0 | 0 |

*41,3*
- 13,7
27,6

fire-log of 10 cm

| | ]0-1.5] | ]1.5-5] | ]5-10] |
|---|---|---|---|
| 37% | 0,7 | 8,5 | **27,6** |

*13,6*
- 4,3
- 8,5
0,8

fire-log of 5 cm

| | ]0-1.5] | ]1.5-5] |
|---|---|---|
| 1% | 0,1 | 0,8 |

0,7
0,7
*1,5% lacking*

fire-log of 1.5 cm

| | ]0-1.5] |
|---|---|
| 0% | 0,0 |

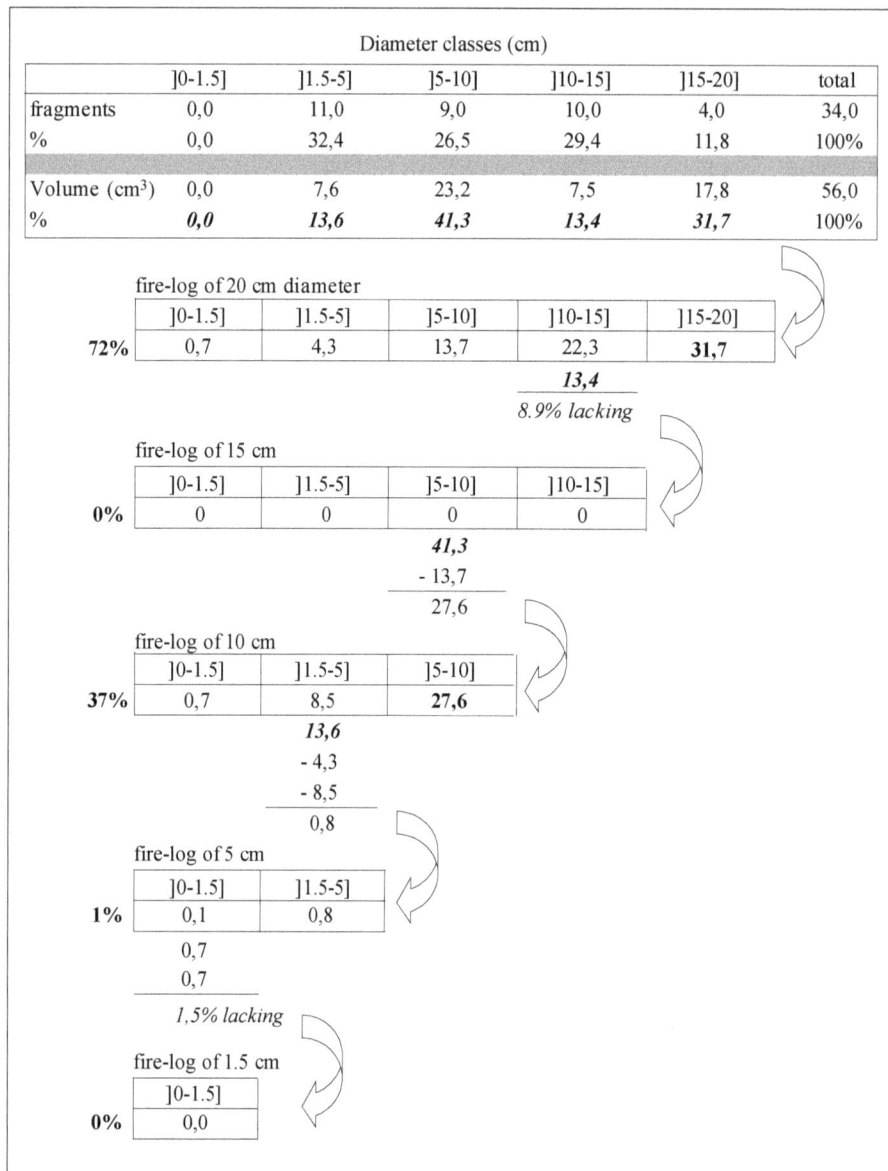

**Figure 7.** This scheme shows the different steps from the relative distribution of the diameter classes in an archaeological sample to the reconstruction of the relative distribution of the trunk diameter that correspond by using the "model table".

the archaeological sample, whereas the larger classes are represented. Indeed, we should have at least 1.5% of the smallest class due to the exploitation of greater diameters. Furthermore, this is in contradiction with the fact that charcoal analysers most often think that the largest classes are under-represented due to the combustion process.

In the figure 9b, we can compare the diameter classes represented in the archaeological sample and the theoretical distribution of the corresponding wood diameter. Whereas four diameter classes are represented, only two sizes of wood diameter were exploited: 10 and 20 cm in diameter. Thus, this model allows greater precision and better interpretation of the results.

## Limits of the model

This model does not take into account criteria such as the behaviour of different wood species during carbonisation, fireplace management, which is unknown, wood morphology (full fire-log or split pieces), etc.

Thus we tested this model with an experiment. This model allows good estimation of the diameter percentages used. Other experiments by Ludemann (in this volume) support the validity of this model. Experimentation has now begun to test this model according to species, diameter, wood morphology (full fire-logs, splints, etc.) and hearth type.

In conclusion, the model provides a good estimation of

| Diameter classes/ fire-log diameter | Distribution of the diameter classes (after carbonisation) | | | Distribution of the theoretical wood diameter (before the carbonisation) | |
|---|---|---|---|---|---|
| | Sample % | theoretical % | corrected % | theoretical % | corrected % |
| ]0-1.5] / 1,5 cm | 0.00 | 1.50 | 1.36 | 0 | 0.0 |
| ]1.5-5] / 5 cm | 13.60 | 3.60 | 12.32 | 1 | 0.9 |
| ]5-10] / 10 cm | 41.33 | 41.30 | 37.41 | 37 | 33.6 |
| ]10-15] / 15 cm | 13.39 | 22.30 | 20.20 | 0 | 0.0 |
| ]15-20] / 20 cm | 31.68 | 31.68 | 28.70 | 72 | 65.5 |
| | | 110.4 | 100 | 110 | 100 |

**Figure 8.** Adjustment to 100% of the distribution diameter classes and of the theoretical distribution of the wood diameter.

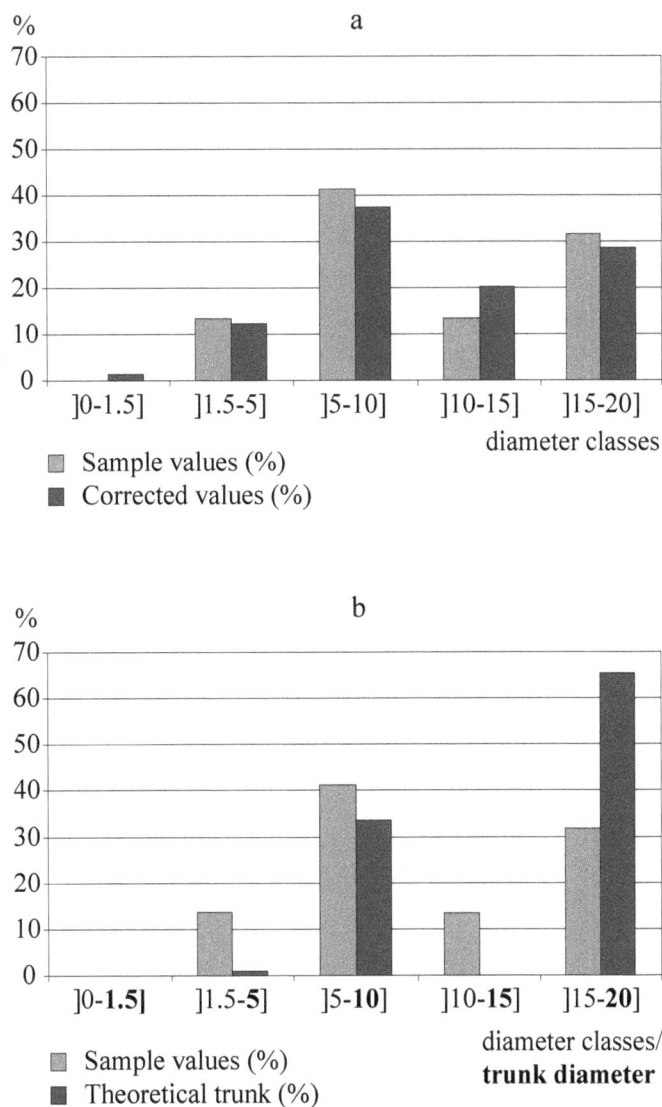

**Figure 9.** a- Comparison between the distribution of the diameter classes in the archaeological sample (light grey) and corrected with the model (dark grey); b- Comparison between the distribution of the diameter classes in the archaeological sample (light grey) and the theoretical distribution of wood diameter which corresponds (dark grey).

wood sizes used by man but, it must be underlined that interpretation in terms of forest stand structure (which depends on species composition, spatial arrangement of trees, etc.) is fastidious chiefly due to the human bias (NELLE 2002). Indeed, whatever the interpretation, it must be remembered that firewood is a resource selected and transported by Man.

## RADIAL GROWTH, AN ECO-ANATOMICAL APPROACH TO DEVELOP

### Basic consideration

We have tried over the last two years to develop an eco-anatomical approach, based on the tree ring-thickness in order to recover the ecological information registered in the tree rings of charcoal. We believe it to be feasible for two reasons. First, radial growth depends on (SCHWEINGRUBER 1996):
- genetic factors (species);
- physiological factors (age, position in the tree...);
- environmental factors (soil, climate, weather, slope, inclination);
- and human factors (clearing, forest management...).

The second reason is found in the palaeoecological representativity of charcoal from an archaeological context, nowadays widely demonstrated (VERNET 1992, THIÉBAULT 2002). Further arguments support this hypothesis: the number of species represented in the charcoal diagrams, the similarity of the results from one site to another (in the same region) and the ecological coherence with the actual vegetation and pollen analysis (CHABAL 1997).

Consequently, analysis of charcoal tree rings should provide information about past environment and human activities (BERNARD ET AL. in this book). However, classical methods used in dendrology (as for dendrochronology, dendroclimtalogy...) cannot be applied to charcoal from archaeological contexts due to their fragmentation. Figure 10 presents, as an example, the distribution of charcoal fragments according to size in a Neolithic lake site. As Chabal (1997) has already described, charcoal fragmentation follows a statistical law, the Poisson law, which is expressed by the best representation of the smallest sizes.

### Laboratory work and paths for data exploitation

The first way consists of measuring tree ring width by means of a reflected light microscope and a micrometer, and to calculate for each fragment, and then for each species in an archaeological layer, an average radial growth. These values can be exploited through statistical tests and then compared from layer to layer in the same site and sometimes between different sites.

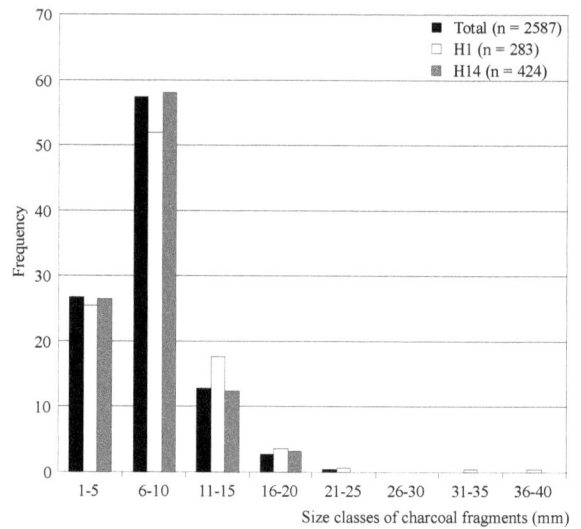

**Figure 10.** Fragmentation of charcoal fragments from Arbon Bleiche 3 (Lake Constance, Switzerland)

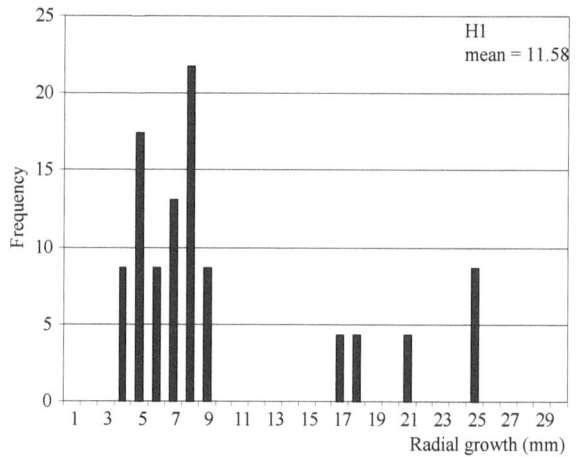

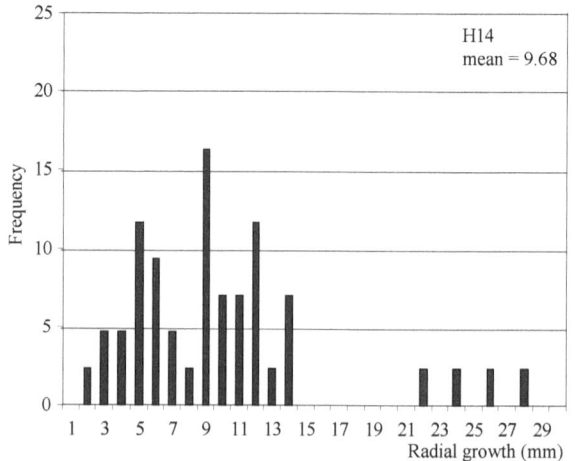

**Figure 11.** Distribution of the radial growth of *Fraxinus* at Arbon Bleiche 3 (Lake Constance, Switzerland)

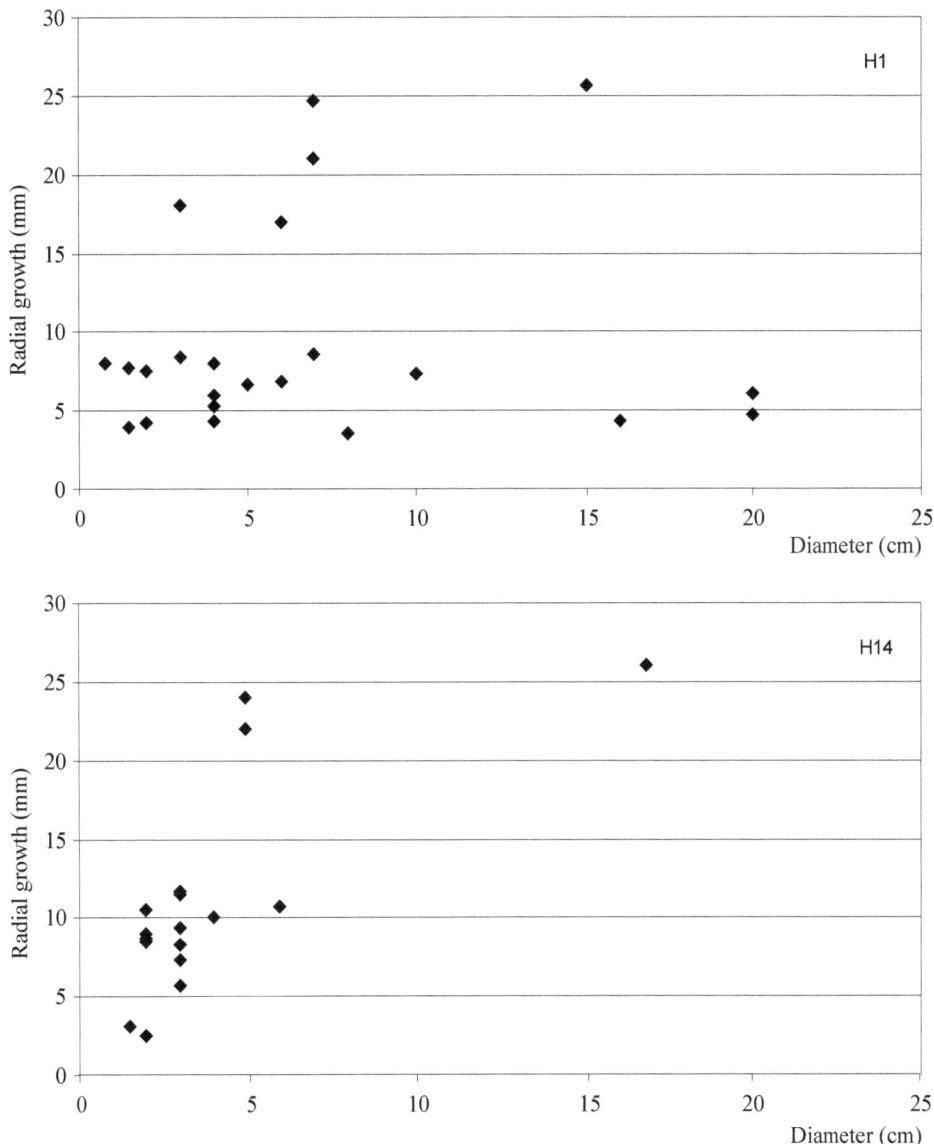

**Figure 12.** Distribution of the radial growth of *Fraxinus* (from H1 and H14 in Arbon Bleiche 3) according to the wood diameter

Figure 11 shows an example from Arbon Bleiche 3. In this site, the percentages of *Fraxinus*, a leader species, vary from house to house. Among the possible hypothesis, we can suppose different acquisition area for each house which can be reflected in the radial growth of *Fraxinus*. This Figure presents a comparison of the *Fraxinus* growth between two houses, H1 and H14. The corpus of *Fraxinus* is composed by 294 fragments and caracterised by 3.9 tree rings per fragment. The two graphs show clearly two distinct population of *Fraxinus*. However, we have to take into account the wood diameter exploited in each house. Thus, the second way is based on the measure of tree-ring width combined with the measure of tree-ring bending. For each species, a growth distribution was then established according to the diameter as is shown by Figure 12 for *Fraxinus*.

Now, how must these distributions be interpreted?

The first observation that we can make is that the studied charcoal fragments in a layer of an archaeological site come from many different trees exploited during the occupation (from 10 to 100 years according to the site) which is why charcoal analysis provides a synthetic image of the former forest. Consequently, these graphs represent the growth of the exploited woodland at a given period as well as its growth conditions. Likewise, it can be considered that all individual variations in a tree, or from tree to tree of the same species, are erased.

The second observation concerns the annual growth of the small diameter classes. Indeed, as we said for the calibre model, one part of the small diameter comes from the centre of large pieces of wood, and the other part from the exploitation of small diameters such as twigs or branches. Besides, we know that the radial growth in a twig, in a branch, in a root or in a trunk is not the same and we must consider these data separately which at least is possible by means of anatomical features. Therefore, the interpretation of such a graph must include wood diameter and the exploited part of the tree (trunk, root, branch…).

The third observation we would like to make concerns the use of current references based on modern forests according to the stand structure, soil type, etc. However, these data are essentially based on commercial forests such as evenage or monospecific forests. Therefore, the greatest trouble is the lack of referential for unmanaged forests which might serve as a comparison with archaeological material. In addition, all measures realised by foresters are always taken at the same height (130 cm) and we cannot specify from which height of the trunk charcoal fragments come from. Another option is to compare carbonised and fresh wood from the same site (from architectural pieces and on which dendrochronological studies were carried out) taking into account shrinking during carbonisation.

Finally, interpretations are based on main events, and only on the relative evolution of radial growth from one layer to another and from one site to another on the same land. Thus, this type of approach provides a new proxy to evaluate past environmental conditions even if it must be further developed.

To conclude, the charcoal analysis is not limited to the identification of wood charcoal and the establishment of a floristic list interpreted in terms of composition and evolution of the forest. The entire exploitation of charcoal, through the study of anatomical features, provides valuable information on palaeoecology, but also on the economic behaviour of Man. Henceforth, all charcoal analysis must take into account these new approaches (see also the other articles in this book) for a new development of anthracology.

---

[1] In fact, a trunk or a twig is a stacking of cones but the sides of the cones form an angle so small that we can assimilate the cones to cylinders. In addition, if we consider cones or cylinders, the relative distribution of the volume for each cylinder (or cone) is the same.

REFERENCES

CHABAL L., 1997.- *Forêts et sociétés en Languedoc (Néolithique final, Antiquité tardive). L'anthracologie, méthode et paléoécologie*, Paris, Editions Maison des Sciences de l'Homme, 189 p. (Documents d'Archéologie Française, 63).

DUFRAISSE A., 2002.- *Les habitats littoraux néolithiques des lacs de Chalain et Clairvaux (Jura, France): collecte du bois de feu, gestion de l'espace forestier et impact sur le couvert arboréen entre 3700 et 2500 av. J.-C. Analyses anthracologiques*. Doctorat, Université de Franche-Comté, 349 p.

DUFRAISSE A., 2005.- Economie du bois de feu et sociétés néolithiques. Analyses anthracologiques appliquées aux sites d'ambiance humide de Chalain et Clairvaux (Jura, France), *Gallia Préhistoire*, 47: 187-233 p.

DUFRAISSE A., PÉTREQUIN A.-M., PÉTREQUIN P., IN PRESS.- La gestion du bois de feu: un undicateur des contextes socio-écologiques. Approche ethno-archéologique dans les Hautes Terres de Papua (Nouvelle-Guinée indonésienne), *Colloque INTERNEO (Neuchâtel, septembre 2005)*.

GREGUSS P., 1959.- *Holzanatomie der Europäischen laubhölzer und sträucher*, Budapest, Akademiai Kiado, 330 p.

JACQUIOT C., 1955.- *Atlas d'anatomie des bois de conifères*, Paris, Centre Technique du Bois, 133 p.

JACQUIOT C., TRENARD Y., DIROL D., 1973.- *Atlas d'anatomie des bois des Angiospermes*, Paris, Centre Technique du Bois, 175 p.

LUDEMANN T., NELLE O., 2002.- *Die Wälder am Schauinsland und ihre Nutzung durch Bergbau und Köhlerei*, Freiburg i.Br., Forstwissenschatliche Fakultät der Universität Freiburg und Forstlische Versuchs- und Forschungsanstalt Baden-Württemberg, 139 p. (Schriftenreihe Freiburger Forstliche Forschung, 15).

MARCONETTO A.B., 2002.- Analysis of burnt building structures of the Ambalo valley (Catamarca, Argentina), in: S. Thiébault (ed.), *Charcoal analysis, methodological approaches palaeoecological results and wood-uses. Proceeding of the second international meeting of anthracology (Paris, september 2000)*, Oxford, BAR Publishing: 267-271 (BAR International Series 1063).

MARGUERIE D., 1992.- *Evolution de la végétation sous l'impact anthropique en Armorique du Mésolithique aux périodes historiques*, Rennes, Editions U.P.R. n° 403 du C.N.R.S., 412 p. (Travaux du Laboratoire d'Anthropologie de Rennes, 40).

NELLE O., 2002.- Charcoal burning remains and forest stand structure. Examples from the Black Forest (south-west Germany) and the Bavarian Forest (south-east Germany), in: S. Thiébault (ed.), *Charcoal analysis, methodological approaches palaeoecological results and wood-uses. Proceeding of the second international meeting of anthracology (Paris, september 2000)*, Oxford, BAR Publishing: 201-207 (BAR International Serie 1063).

SCHWEINGRUBER F. H., 1990a.- *Anatomie europaïscher Hölzer*, Bern/Stuttgart, Verlag Paul Haupt, 800 p.

SCHWEINGRUBER F.H., 1990b.- *Anatomie microscopique du bois*, Birmensdorf, Editions de l'Institut fédéral de recherches sur la forêt, la neige et le paysage, 226 p.

SCHWEINGRUBER F.H., 1996.- *Tree rings and Environnement. Dendroecology,* Bern/Stuttgart/Wien, Verlag Paul Haupt, 609 p.

SCHWEINGRUBER F.H., 2001.- *Dendroökologische Holzanatomie. Anatomische Grunglagen der Dendrochronologie,* Bern/ Stuttgart/Wien, Verlag Paul Haupt, 472p.

SHAKELTON C.M., PRINS F., 1992.- Charcoal Analysis and the "Principe of Least Effort"- A Conceptual Model, *Journal of Archeological Science,* 19: 631-637.

SMART T.L., HOFFMAN E.S., 1987.- Environnemental interpretation of archaeological charcoal, *in*: C.A. Hastorf and V.S. Popper (eds), *Current Paleoethnobotany: analytical methods and cultural interpretation of archaeological plant remains,* Chicago/London, University Chicago Press: 167-205.

THÉRY-PARISOT I., 1998.- *Economie du combustible et paléoécologie en contexte glaciaire et périglaciaire, Paléolithique moyen et supérieur du sud de la France. Anthracologie, expérimentation, taphonomie,* Doctorat, Université de Paris I, 499 p.

THÉRY-PARISOT I., 2001.- *Economie des combustibles au Paléolithique. Expérimentation, taphonomie, anthracologie,* Paris, Editions C.N.R.S. - C.E.P.A.M., 195 p. (Dossier de Documentation Archéologique, 20).

THIÉBAULT S., 2002.- *Charcoal analysis, methodological approaches palaeoecological results and wood-uses. Proceeding of the second international meeting of anthracology (Paris, september 2000),* Oxford, BAR Publishing, 284 p. (BAR International Series 1063)

VERNET J.-L., 1992.- *Les Charbons de bois, les anciens écosystèmes et le rôle de l'homme. Colloque organisé à Montpellier en décembre 1991,* Paris, Editions du Bulletin de la Société Botanique Française, 725 p. (Actualités Botaniques 1992-2/3/4).

# ANTHRACOLOGICAL ANALYSIS OF RECENT CHARCOAL-BURNING
# IN THE BLACK FOREST, SW GERMANY

THOMAS LUDEMANN

University of Freiburg, Institute of Biology II, Dept. of Geobotany
Schaenzlestrasse 1, D-79104 Freiburg (Germany)
thomas.ludemann@biologie.uni-freiburg.de

ABSTRACT: We have developed a standardized anthracological method with special regard to the diameter of the wood used. Thereby the charcoal fragments were sized by the curvature of the tree rings and by the angles of the rays. Based on diameter size-class distributions, mean diameter values were calculated. The specific distributions and diameter values could be used for the anthracological characterization of historic charcoal samples. Information on the proportions and dimensions of the wood taxa exploited in the past was deduced. In order to verify the method and the interpretation of the results, remains from recent charcoal-burning were analysed. Hereby, the wood used was known, and its theoretical distribution to diameter size-classes was calculated. After finishing the charcoal-burning, charcoal samples were taken in the same way as at the ancient sites. The wood used and the anthracological results were compared. The specific characteristics of the fuel and construction wood actually used for charcoal-burning at the recent kiln site were documented well by the remains and their anthracological analysis. The quantitative differences in proportions and dimensions of the tree taxa used could be deduced from the anthracological results. The analysis of recent charcoal-burning indicates that the method applied for historic samples is a sensitive tool to establish qualitative and quantitative information about the wood taxa compositions and diameters used in past charcoal production.

KEY WORDS: Anthracological diameter analysis, Charcoal-burning, Experimental archaeology, Fuel wood, Kiln site anthracology, Recent charcoal analysis

RÉSUMÉ: Nous avons développé une méthode standard pour restituer, à partir des charbons de bois, les diamètres de bois utilisé dans le passé. Les diamètres des charbons de bois ont été mesurés par les courbures de cerne et l'angle des rayons. Les diamètres moyens ont été calculés sur la base de la distribution des différentes classes de diamètre. La distribution spécifique et les valeurs de diamètre permettent de caractériser les échantillons de charbons historiques et d'en déduire des informations sur les proportions entre essences et les dimensions des bois exploités. Pour vérifier cette méthode et l'interprétation des résultats, les bois carbonisés issus de charbonnières actuelles ont été analysés; le bois utilisé était connu au départ, ce qui a permi de calculer la distribution théorique des diamètres. La combustion une fois terminée, les échantillons ont été prélevés selon les mêmes techniques qu'en contexte historique. Puis, les résultats de l'analyse anthracologique ont été comparés aux données de départ. Les différences quantitatives dans les proportions et les dimensions des trois essences utilisées ont pu être retrouvées par l'analyse anthracologique. Cette expérimentation montre que cette méthode est un bon outil qui apporte des informations qualitatives et quantitatives sur la composition et les diamètres de bois utilisé dans le passé en contexte de charbonnage *(Translation Alexa Dufraisse)*.

MOTS CLÉS : calibre de bois, charbonnières, archéologie expérimentale, combustible, analyse de charbons actuels

ZUSAMMENFASSUNG: Wir haben eine einheitliche anthrakologische Analysemethode unter besonderer Berücksichtigung der Stärke des verwendeten Holzes entwickelt. Die Holzkohlestücke werden dabei anhand der Jahrringkrümmung und des Winkels der Markstrahlen verschiedenen Durchmesserklassen zugeordnet. Die ermittelten Verteilungen auf die Durchmesserklassen und die daraus errechneten Mittelwerte werden zur Charakterisierung von historischen Holzkohleproben verwendet und davon Aussagen zu Anteil und Stärke der in der Vergangenheit genutzten Holzarten abgeleitet. Um die Methode zu überprüfen und die Interpretation der Ergebnisse abzusichern, wurden Rückstände der Holzkohle-Herstellung untersucht, von denen das verwendete Ausgangsmaterial bekannt war. Die charakteristischen Eigenschaften des genutzten Holzes spiegeln sich in den Rückständen und den holzkohleanalytischen Ergebnissen klar wider. Die wesentlichen Unterschiede in Anteil und Stärke der verwendeten Holzarten lassen sich gut ableiten. Die rezent-anthrakologischen Untersuchungen belegen, dass mit der entwickelten Methode auch anhand der historischen Rückstände qualitative

und quantitative Aussagen zur Baumartenzusammensetzung und Stärke des früher verwendeten Holzes erzielt werden können.
STICHWORTE: Anthrakologie, Durchmesseranalyse, Energieholznutzung, experimentelle Archäologie, Holzkohleherstellung, Köhlerei, Rezent-Holzkohleanalyse

## INTRODUCTION

For several years remains of historic charcoal-burning in the Black Forest have been analysed systematically. A standardized anthracological method was developed, with special regard to the diameters of wood exploited in the past. Diameter size-class distributions and mean diameter values were determined for most of the charcoal samples. This method was applied in many studies and for large numbers of charcoal remains (LUDEMANN 1995, 1996, 1999, 2001, LUDEMANN AND BRITSCH 1997, LUDEMANN AND NELLE 2002, NELLE 2002a, b, NÖLKEN 2003, 2004, 2005). From anthracological analyses we attempt to deduce information on the taxa compositions and dimensions of the exploited wood.

But, in fact, we do not know the real compositions and dimensions of the wood from which the charcoal remains come. Up to now the calculated mean diameter values were only relative values to compare different samples. The calculation was done in order to simplify the comparison of samples and to describe the relation of the wood dimension of different taxa and samples within an order of magnitude.

Indications of the relation of the diameter of the used wood on the one hand and anthracologically determined mean diameter on the other were given by Ludemann (1996), Ludemann and Nelle (2002) and Nelle (2002a, b). In addition, in the last few years we started to study remains of recent and experimental charcoal-burning, whereby the used wood was known, in order to verify the interpretations of the historic material. We want to verify the quantitative relation between anthracologically determined size-class distributions and mean diameter values on the one hand and the actually used wood diameters on the other.

In this study the currently managed kiln site of one of the last professional German charcoal burners was analysed by the same standardized method, as was applied at the historic sites. From discussions with the charcoal burner and by building-up a kiln together with him, we came to know which wood he used for charcoal production and kiln construction. Based on this information, the size-class distributions and mean diameter values were calculated for the charcoal theoretically arising from the wood used. These calculations were compared with the anthracological results of the recent charcoal-burning site.

**Figure 1.** The studied kiln site with a burned finished charcoal kiln, sheltered by a roof construction, and bags with charcoal prepared for sale. Münstertal-Gabel, Black Forest, SW Germany. 7 June 2000.

## STUDY SITE

We analysed the kiln site of the last professional charcoal burner in the southwestern part of the Black Forest. This kiln site is located in a small remote valley neighbouring the village of Münstertal in the Black Forest close to the Rhine Valley. The currently managed site is characterized by a deep black soil layer. Within this layer charcoal remains were accumulated of all sizes which were too small for sale. The charcoal burner was descended from an old charcoal burner family. He learned his special art (charcoal-burning in upright circular kilns) in the traditional historic manner from his uncle and has burned charcoal for thirty years. Figures 1 to 3 show the recent kiln site studied; Figure 4 schematically the kiln construction. The charcoal burner has covered his working site with a roof construction in order to be able to work under dry conditions independently of the weather.

## MATERIAL AND METHODS

### The wood used

Information on the wood used, the kiln construction and the charcoal-burning process were given both (1) by the charcoal burner himself, answering all our questions, and (2) by personal observations and experience during site visits and particularly in building-up a kiln together with the charcoal burner.

## Diameter distribution - calculation for charred wood

In order to link anthracological results and wood use, the theoretical volume distribution of the used wood to diameter size-classes I-V (see FIG. 5) was calculated, based on the following calculation and mathematical formulas for cylinders, cones and parts of cones:

Cylinder volume: $Vol\_cyl = \pi(d/2)(d/2)l$

Cone volume: $Vol\_cone = Vol\_cyl/3 = \pi(d/2)(d/2)l/3$

Cone part volume: $Vol\_pcone = \pi(r_1*r_1+r_1*r_2+r_2*r_2)l/3$

with d, diameter; l, length of cylinder, cone or cone part; $r_1$ and $r_2$, minimum and maximum radius of cone part.

The distinguished diameter size-classes correspond to those distinguished in anthracological wood diameter determination (cf. section on anthracological analysis). The calculated volume distributions for wood of specific diameters are given in Figure 6. The same calculations were done for (theoretically) charred wood (FIG. 7 and 8). In the latter calculations, the reduction of diameter of about 20%, caused by radial and tangential shrinking of wood during carbonisation (BROCKHAUS 1931, 2001, HERDER 1954, SCHLÄPFER AND BROWN 1948, SCHWEINGRUBER 1990), was taken into account.

**Figure 2.** Kiln in construction in the background to the right with large fuel wood ("Scheitholz", diameter > 14 cm). Left hand side, small fuel wood ("Prügelholz", diameter 7-14 cm) and cover material (*Picea abies* twigs). 9 June 2001.

**Figure 3.** A burning kiln. Münstertal-Gabel, Black Forest, SW Germany. 25 July 1996.

**Figure 4.** Construction of an upright circular kiln. 1 Kiln base, "Rost". 2 Kiln centre, "Quandel". 3 Fuel wood, "Kohlholz"; *Fagus, Acer,* etc trunk and branch wood. 4 Green cover/roof, "Gründach"; *Picea* and *Abies* twigs. 5 Black or soil cover/roof, "Schwarzdach/ Erddach" (Drawing by NELLE 1998, from LUDEMANN AND NELLE 2002, modified).

## Sampling of charcoal remains

A sampling strategy most similar to that used at the historic kiln sites was applied at the recent kiln site. Charcoal fragments of sizes larger than 0.25 cm³ (corresponding to a mesh width of 5 mm) were collected by hand from all over the charcoal layer at or close to the soil surface. More than 200 charcoal fragments were collected and put together to make one mixed charcoal sample.

## Anthracological analysis

The determination of wood taxa of the charcoal fragments followed Schweingruber (1990), using a stereoscope (Leica MZ 12) and an incident-light microscope (Zeiss Universal M III C) as well as a reference collection of charred known wood. The quantity of each established taxon was determined in two ways, by both the number of pieces counted and their weight. Altogether, 222 charcoal fragments with a total weight of 385 g were analysed.

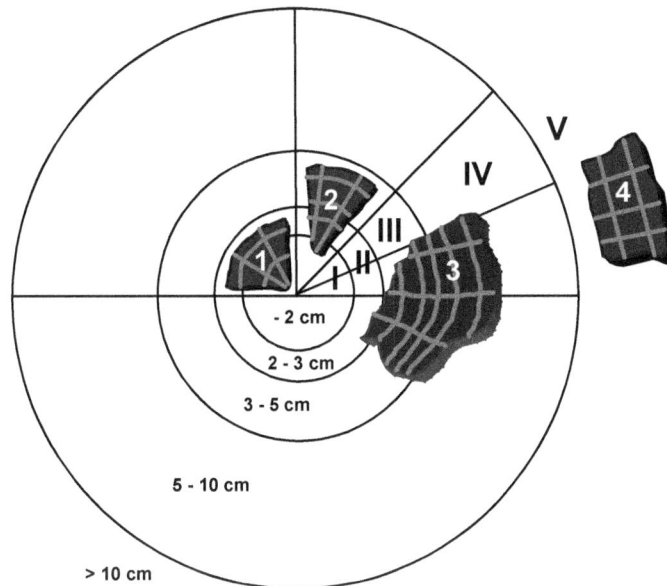

**Figure 5.** Diameter template, diameter size-classes (I-V) and examples of charcoal fragments to be sized (1-4). Result of sizing: Fragment 1, class II; fragment 2, class III; fragment 3, class IV; fragment 4, class V.

For diameter determination the charcoal fragments were sized by the curvature of the annual growth rings and by the angles of the rays to each other, using a diameter template (FIG. 5). Five wood diameter classes were distinguished; smaller than 2 cm, 2 to 3 cm, 3 to 5 cm, 5 to 10 cm, and larger than 10 cm. In this way diameter size-class distributions for the samples and the taxa were obtained.

Based on the diameter size-class distribution, a single value is calculated, termed the mean diameter $mD = (n_I + n_{II}*2.5 + n_{III}*4 + n_{IV}*7.5 + n_V*15)/N$; where $n_I$ to $n_V$ is the number of fragments in the respective class, and N the total number of fragments analysed from the sample. The mean diameter can theoretically vary between 1 and 15 cm if all fragments are in the first class or all in the fifth class.

Indeed, successful determination requires charcoal samples with a sufficient number of fragments large enough and suited wood-anatomically for this kind of sizing. 210 charcoal fragments could be sized in this way; for 12 charcoal pieces the determination of a size-class was not possible.

## RESULTS

### Wood used for the kilns

Trunk wood of *Fagus sylvatica* was mainly used as fuel wood, in addition those of *Acer pseudoplatanus* and other indigenous deciduous wood species present in the forests of the surroundings of the charcoal-burning site; e.g.

*Alnus glutinosa, Carpinus betulus, Corylus avellana* and *Fraxinus excelsior.* Moreover, conifer wood of *Abies alba* and *Picea abies* were used; logs of conifer trunk wood have been situated preferentially in the centre of the kilns, in order to guarantee a good start of the charcoal-burning process.

About 90% of the fuel wood used for charcoal-burning had diameters of more than 14 cm - to 30 or 40 cm as a maximum (trunk sections/cylinders; in German: "Scheitholz"). That is the usual dimension of recent fire wood collections exploited and sold today from the communal forests of Münstertal, in which the studied kiln site was located. About 10% had diameters between 7 and 14 cm (in German: "Prügelholz"). Wood with larger diameters (> 14 cm) was mainly of *Fagus, Acer, Abies* and *Picea*, while the proportions of "Prügelholz" – diameters 7 to 14 cm – were larger for the other deciduous species, especially of shrubs like *Corylus avellana* and *Sambucus racemosa* and pioneer species like *Betula pendula* and *Salix caprea*. The cover material of the kilns consisted of twigs of *Picea abies* – which was preferred – and of *Abies alba*.

### Size-class distribution and mean diameter of charred wood

The (theoretical) size-class distributions and mD-values of the used wood dimensions are given in figures 6, 7 and 8. Most of the charred wood of 7 cm diameter cylinders (minimum size of "Prügelholz") after carbonisation belong to the third diameter size-class (FIG. 8); those

| Wood diameter (cm) | | Volume (%) of wood in Diameter size-classes | | | | | mD Wood (cm) |
|---|---|---|---|---|---|---|---|
| | | I | II | III | IV | V | |
| 3.0* | | 74.1 | 25.9 | 0.0 | 0.0 | 0.0 | 1.4 |
| 7.0 | | 8.2 | 10.2 | 32.7 | 49.0 | 0.0 | 5.3 |
| 10.5 | (7-14 cm) | 3.6 | 4.5 | 14.5 | 68.0 | 9.3 | 7.2 |
| 14.0 | | 2.0 | 2.6 | 8.2 | 38.3 | 49.0 | 10.6 |
| 17.0 | | 1.4 | 1.7 | 5.5 | 26.0 | 65.4 | 12.0 |
| 22.0 | (14-30 cm) | 0.8 | 1.0 | 3.3 | 15.5 | 79.3 | 13.2 |
| 27.0 | (14-40 cm) | 0.5 | 0.7 | 2.2 | 10.3 | 86.3 | 13.8 |
| 30.0 | | 0.4 | 0.6 | 1.8 | 8.3 | 88.9 | 14.0 |
| 40.0 | | 0.3 | 0.3 | 1.0 | 4.7 | 93.8 | 14.5 |

**Figure 6**. Volume distribution of wood to diameter size-classes I to V. Calculated for wood cylinders of a specific diameter. Diameter size-class I: 0-2 cm; II: >2-3 cm; III: >3-5 cm; IV: >5-10 cm; V: >10 cm. mD mean diameter value (cf. section on anthracological analysis). *Calculated for cones.

| Diameter | | Volume (%) of charred wood/charcoal in Diameter size-classes | | | | | mD Charred |
|---|---|---|---|---|---|---|---|
| Wood (cm) | Charred** wood (cm) | I | II | III | IV | V | wood (cm) |
| 3.0* | 2.4 | 92.6 | 7.4 | 0.0 | 0.0 | 0.0 | 1.1 |
| 7.0 | 5.6 | 12.8 | 15.9 | 51.0 | 20.3 | 0.0 | 4.1 |
| 10.5 | 8.4 | 5.7 | 7.1 | 22.7 | 64.6 | 0.0 | 6.0 |
| 14.0 | 11.2 | 3.2 | 4.0 | 12.8 | 59.8 | 20.3 | 8.2 |
| 17.0 | 13.6 | 2.2 | 2.7 | 8.7 | 40.5 | 45.9 | 10.4 |
| 22.0 | 17.6 | 1.3 | 1.6 | 5.2 | 24.2 | 67.7 | 12.2 |
| 27.0 | 21.6 | 0.9 | 1.1 | 3.4 | 16.1 | 78.6 | 13.2 |
| 30.0 | 24.0 | 0.7 | 0.9 | 2.8 | 13.0 | 82.6 | 13.5 |
| 40.0 | 32.0 | 0.4 | 0.5 | 1.6 | 7.3 | 90.2 | 14.2 |

**Figure 7.** Volume distribution of charred wood/charcoal to diameter size-classes I to V. Calculated for the wood cylinders of Figure 6 after carbonisation (**diameter reduction; radial and tangential shrinking about 20% taken into account). Diameter size-class I: 0-2 cm; II: >2-3 cm; III: >3-5 cm; IV: >5-10 cm; V: >10 cm. mD mean diameter value (cf. section on anthracological analysis). *Calculated for cones.

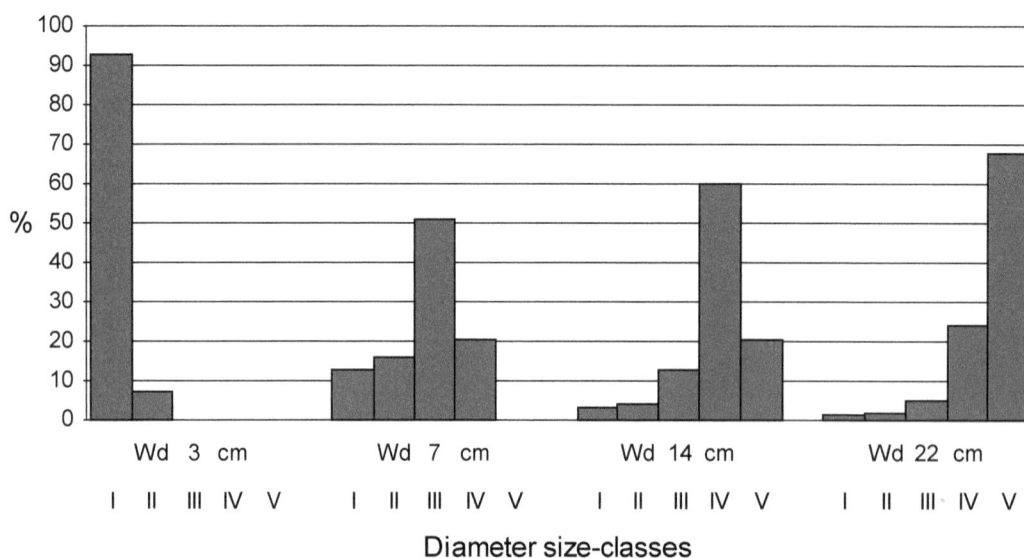

**Figure 8.** Distribution of charcoal volume of a specific wood diameter (Wd) to diameter size-classes (after charbonisation; cf. Fig. 7). Calculated for wood cylinders (Wd 7, 14, 22 cm) and for wood cones (Wd 3 cm), taking volume reduction by carbonisation into account. Diameter size-class I: 0-2 cm; II: >2-3 cm; III: >3-5 cm; IV: >5-10 cm; V: >10 cm.

Anthracological Analysis of Recent Charcoal-Burning in the Black Forest

|  | N % | W % | Distribution of charcoal pieces n to Diameter size-classes | | | | | mD Charcoal (cm) |
|---|---|---|---|---|---|---|---|---|
|  |  |  | I | II | III | IV | V |  |
| *Fagus* | 50.9 | 58.5 |  |  | 2 | 31 | 75 | 12.6 |
| *Picea* | 10.4 | 12.8 | 18 |  | 1 | 1 | 3 | 3.2 |
| *Abies* | 9.5 | 6.6 | 8 |  |  |  | 13 | 9.7 |
| *Acer* | 9.5 | 4.9 |  |  | 5 | 4 | 11 | 10.8 |
| *Corylus* | 5.4 | 3.7 |  |  | 5 | 4 | 1 | 6.5 |
| *Alnus* | 4.5 | 3.1 |  |  | 1 | 5 | 2 | 8.9 |
| *Fraxinus* | 2.7 | 2.5 |  | 1 | 2 | 2 | 1 | 6.8 |
| *Carpinus* | 2.3 | 4.0 |  |  |  | 5 |  | 7.5 |
| *Sambucus* | 2.3 | 1.6 |  |  |  | 4 |  | 7.5 |
| *Betula* | 1.8 | 1.4 |  |  | 1 | 2 |  | 6.3 |
| *Salix* | 0.5 | 0.6 |  |  |  | 1 |  | 7.5 |
| *Quercus* | 0.5 | 0.4 |  |  |  | 1 |  | 7.5 |
| Total N | 222 |  | 26 | 1 | 17 | 60 | 106 | 10.2 |

**Figure 9.** Frequency distribution of charcoal fragments (N) and weight (W) to taxa and diameter size-classes I to V. Diameter size-class I: 0-2 cm; II: >2-3 cm; III: >3-5 cm; IV: >5-10 cm; V: >10 cm. mD mean diameter value (cf. section on anthracological analysis).

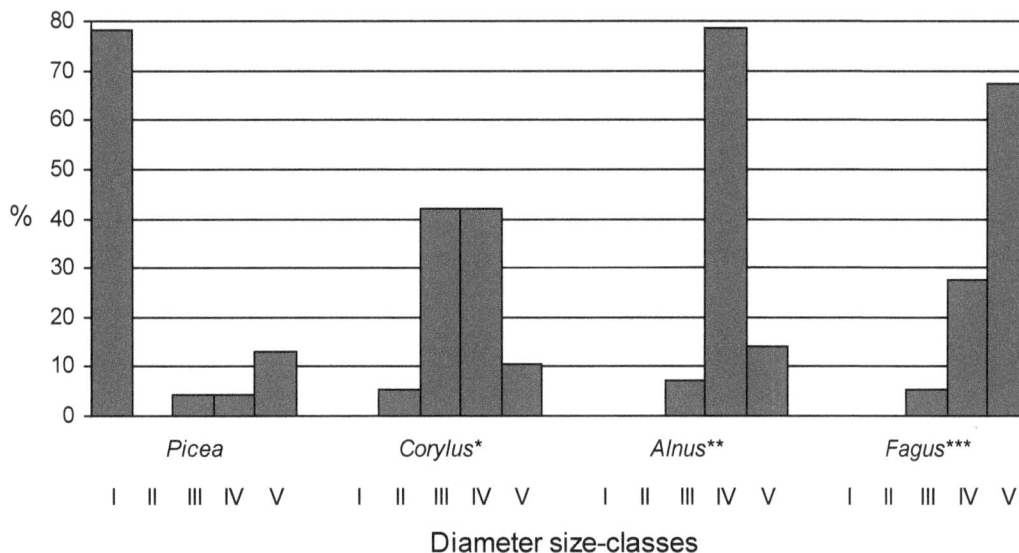

**Figure 10.** Distribution of charcoal fragments of specific taxa to diameter size-classes. Diameter size-class I: 0-2 cm; II: >2-3 cm; III: >3-5 cm; IV: >5-10 cm; V: >10 cm. *+*Fraxinus+Betula*; **+*Carpinus+Quercus*; ***+*Acer*.

of 10.5 cm diameter (medium size of "Prügelholz") as well as those of 14 cm diameter (maximum size of "Prügelholz" and minimum size of "Scheitholz") have the largest volume proportions in the fourth class (5-10 cm diameter; FIG. 7 and 8). The other (larger) wood diameters calculated ("Scheitholz" ≥ 17 cm) have a maximum in the largest size-class. The mean diameter mD of charred "Prügelholz" lie between 4.1 and 8.2 cm. The mD-value of charred "Scheitholz" is more than 8.2 cm. The volume distributions of wood of 3 cm diameter were calculated for cones, considering the material used to cover the kilns (*Picea* and *Abies* twigs). A very large proportion of this wood material falls to the smallest size-class and the mD-value is correspondingly small.

## Charcoal remains

In the charcoal material analysed (222 fragments) we found 12 wood taxa. These were, with decreasing frequency, *Fagus, Picea, Abies, Acer, Corylus, Alnus, Fraxinus, Carpinus, Sambucus, Betula, Salix* and *Quercus*. More than half of the material belonged to *Fagus*. Second place was taken by *Acer, Abies* and *Picea*, with similar

| | Diameter (cm) | Distribution (%) to Diameter size-classes | | | | | mD (cm) |
|---|---|---|---|---|---|---|---|
| | | I | II | III | IV | V | |
| Charred wood | 3 | 93 | 7 | 0 | 0 | 0 | 1.1 |
| *Picea* charcoal | <3 (>14) | 78 | 0 | 4 | 4 | 13 | 3.2 |
| Charred wood | 7 | 13 | 16 | 51 | 20 | 0 | 4.1 |
| Charred wood | 10.5 (7-14) | 6 | 7 | 23 | 65 | 0 | 6.0 |
| *Betula* charcoal | 7-14 | 0 | 0 | 33 | 67 | 0 | 6.3 |
| *Corylus* charcoal | 7-14 | 0 | 0 | 50 | 40 | 10 | 6.5 |
| *Fraxinus* charcoal | 7-14 | 0 | 17 | 33 | 33 | 17 | 6.8 |
| *Carpinus* charcoal | 7-14 | 0 | 0 | 0 | 100 | 0 | 7.5 |
| *Sambucus* charcoal | 7-14 | 0 | 0 | 0 | 100 | 0 | 7.5 |
| *Salix* charcoal | 7-14 | 0 | 0 | 0 | 100 | 0 | 7.5 |
| *Quercus* charcoal | 7-14 | 0 | 0 | 0 | 100 | 0 | 7.5 |
| Charred wood | 14 | 3 | 4 | 13 | 60 | 20 | 8.2 |
| *Alnus* charcoal | 7-14 | 0 | 0 | 13 | 63 | 25 | 8.9 |
| *Abies* charcoal | >14 (<3) | 38 | 0 | 0 | 0 | 62 | 9.7 |
| Total charcoal | >14 (>7, <3) | 12 | 0 | 8 | 29 | 50 | 10.2 |
| Charred wood | 17 | 2 | 3 | 9 | 41 | 46 | 10.4 |
| *Acer* charcoal | >14 | 0 | 0 | 25 | 20 | 55 | 10.8 |
| Charred wood | 22 (14-30) | 1 | 2 | 5 | 24 | 68 | 12.2 |
| *Fagus* charcoal | >14 | 0 | 0 | 2 | 29 | 69 | 12.6 |
| Charred wood | 27 (14-40) | 1 | 1 | 3 | 16 | 79 | 13.2 |
| Charred wood | 30 | 1 | 1 | 3 | 13 | 83 | 13.5 |
| Charred wood | 40 | 0 | 0 | 2 | 7 | 90 | 14.2 |

**Figure 11.** Distribution of charcoal fragments (grey fields) and charred wood of a specific diameter (white fields) in diameter size-classes I to V. Diameter size-class I: 0-2 cm; II: >2-3 cm; III: >3-5 cm; IV: >5-10 cm; V: >10 cm. mD mean diameter value (cf. section on anthracological analysis).

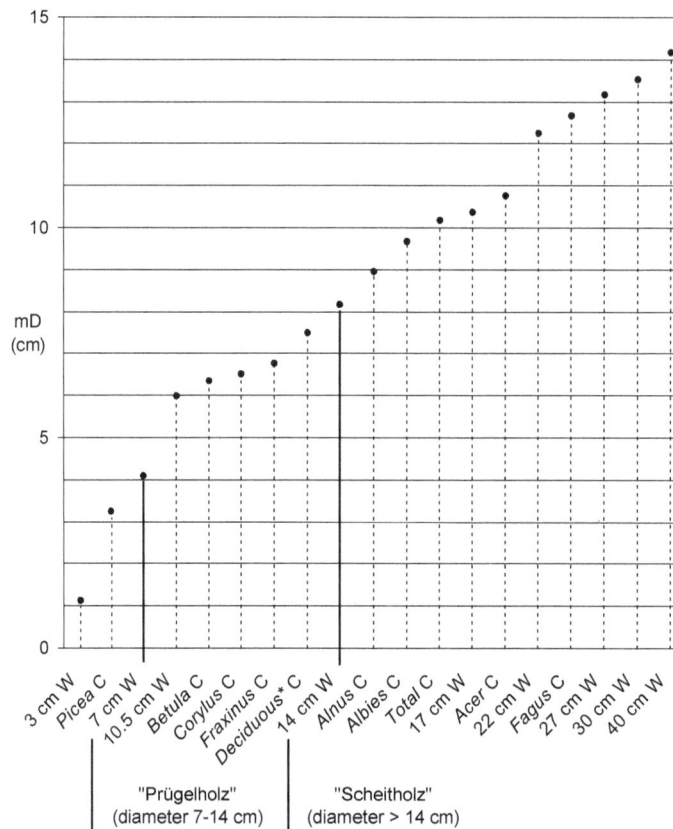

**Figure 12.** mD-values calculated for wood of a specific diameter (W) and determined anthracologically for charcoal remains of specific taxa (C). mD: mean diameter value (cf. section on anthracological analysis). *Carpinus, Quercus, Salix, Sambucus*.

proportions of about 10% each. Additionaly, proportions of about 5% were established for *Corylus* and *Alnus*. Half of the taxa attained a proportion less than 3% (FIG. 9).

Considering the order of proportions, the results of the determination based on the number of pieces were quite similar to the results established by weight analysis. We found only one moderate exception; *Carpinus* took place five considering the weight (4%) and place eight following the number of charcoal pieces (2.3%). This is caused by a single exceptionally large charcoal piece of *Carpinus* found at random.

Considering the size-class distributions and mD-values, *Fagus, Acer* and *Abies* assumed the highest values, about ten or more cm, with each having a maximum of pieces in the largest class. Most of the other deciduous wood taxa had similar mD-values - about seven - with most fragments in the fourth or third classes. Pieces of both the smallest and the largest size-classes were found for *Picea* and *Abies*. Most charcoal fragments of *Picea* belong to the smallest class, so that it had the smallest mD-value.

## Charred wood calculation and anthracological result

For comparison in Figure 11 the diameter size-class distributions - calculated both for charred wood of a specific diameter and determined for the charcoal fragments of a specific taxon - were arranged in the order of their magnitude, considering the mD-values.

For the taxa for which mainly "Scheitholz" was used (*Acer, Fagus*), size-class distributions and mD-values characteristic for the corresponding diameters of wood were also established anthracologically (mD 10-13 cm). Similarly, most of the taxa for which mainly "Prügelholz" was used lay between the minimum and maximum of the corresponding wood dimensions (diameters 7-14 cm; mD 6-8 cm; FIG. 11 and 12).

## DISCUSSION AND CONCLUSION

Wood use and anthracological results. The most important information about fuel and construction wood used for the kilns was documented by the charcoal remains and their analysis. Valuable additional information can be provided especially by anthracological diameter analysis. The specific characteristics and differences in proportions and dimensions of the individual wood taxa used could be clearly deduced from the anthracological results:
- mainly large deciduous wood of *Fagus*, but also *Acer, Abies* and a little bit of *Picea* (FIG. 9 and 10, right columns);
- other deciduous species, mainly with smaller diameters (FIG. 10, central columns);
- smallest diameters, mainly of *Picea*, but also of *Abies* (cover material of the kilns; FIG. 9 and 10, left columns).

Moreover, we can deduce a first quantitative relation of the used wood diameters and the mD-values. The use of "Scheitholz" (diameter > 14 cm) may be characterized anthracologically by mD-values of 8 cm and larger, the use of "Prügelholz" (diameter 7 to 14 cm) by mD-values of 4 to 8 cm. Obviously, the use of small branches and twigs is recorded by fragments of the smallest diameter size-class and correspondingly, by very small mD-values.

But these could be only preliminary results, since we could not measure the fuel and construction wood used for charcoal-burning at this kiln site in detail. Furthermore, the number of analysed fragments should be enlarged. Nevertheless, a good and sufficient correlation seems to be indicated. Further investigations should be undertaken, especially studies of experimental charcoal-burning, whereby the used wood is known and measured in detail before burning, in order to compare the results of wood measurement and of anthracology quantitatively (LUDEMANN IN PRESS). These studies should be undertaken at virgin sites without any charcoal remains of former burning events in the soil.

ACKNOWLEDGMENTS

We are grateful to the charcoal burner S. Riesterer for answering all our questions patiently and for giving us the chance to help in constructing a kiln and to participate in the charcoal-burning process. Moreover, we thank R. Cassada (University of Freiburg) for linguistic editing.

REFERENCES

BROCKHAUS, 1931.- *Der Große Brockhaus. Handbuch des Wissens in 20 Bänden*, Leipzig, Brockhaus, 784 p.

BROCKHAUS, 2001.- *Brockhaus. Die Enzyklopädie in 24 Bänden*, Leipzig, Brockhaus, 24 vol.

HERDER, 1954.- *Der Große Herder. Nachschlagwerk für Wissen*

*und Leben*, Freiburg, Herder, 10 vol.

LUDEMANN T., 1995.- Zwei Kohlplätze im Mittleren Schwarzwald, *Mitteilungen badischer Landesverein Naturkunde Naturschutz*, N.F. 16 (2): 319-334.

LUDEMANN T., 1996.- Die Wälder im Sulzbachtal (Südwest-

Schwarzwald) und ihre Nutzung durch Bergbau und Köhlerei, *Mitt. Ver. forstl. Standortskd. Forstpflanzenzücht.*, 38: 87-118.

LUDEMANN T., 1999.- Zur Brennstoffversorgung einer römischen Siedlung im Schwarzwald, *in:* S. Brather, C. Bücker and M. Hoeper (eds.), *Archäologie als Sozialgeschichte. Studien zu Siedlung, Wirtschaft und Gesellschaft im frühgeschichtlichen Mitteleuropa, Festschrift für Heiko Steuer zum 60. Geburtstag,* Leidorf, Rahden/Westfalen: 165-172 (Internationale Archäologie. Studia honoraria, 9).

LUDEMANN T., 2001.- Das Waldbild des Hohen Schwarzwaldes im Mittelalter, Ergebnisse neuer holzkohleanalytischer und vegetationskundlicher Untersuchungen, *Alemannisches Jahrbuch*, 1999/2000: 43-64.

LUDEMANN T., IN PRESS.- *Experimental charcoal-burning with special regard to anthracological wood diameter analysis.* Oxford, BAR Publishing (BAR International Series).

LUDEMANN T., BRITSCH T., 1997.- Wald und Köhlerei im nördlichen Feldberggebiet, Südschwarzwald, *Mitteilungen badischer Landesverein Naturkunde Naturschutz*, N.F. 16 (3-4): 487-526.

LUDEMANN T., NELLE O., 2002.- *Die Wälder am Schauinsland und ihre Nutzung durch Bergbau und Köhlerei,* Freiburg i.Br., Forstwissenschatliche Fakultät der Universität Freiburg und Forstliche Versuchs- und Forschungsanstalt Baden-Württemberg, 139 p. (Schriftenreihe Freiburger Forstliche Forschung, 15).

NELLE O., 1998.- *Waldstandorte und Köhlerei am Schauinsland (Südschwarzwald).* PhD Thesis, Universität Freiburg, 96 p.

NELLE O., 2002a.- Charcoal burning remains and forest stand structure - Examples from the Black Forest (south-west Germany) and the Bavarian Forest (south-east Germany), *in:* S. Thiébault (ed.), *Charcoal analysis, methodological approaches, palaeoecological studies and wood uses, Proceeding of the second international meeting of anthracology (Paris, september 2000)*, Oxford, BAR Publishing: 201-207 (BAR International Series 1063).

NELLE O., 2002b.- Zur holozänen Vegetations- und Waldnutzungsgeschichte des Vorderen Bayerischen Waldes anhand von Pollen- und Holzkohleanalysen, *Hoppea*, 63: 161-361.

NÖLKEN W., 2003.- Holzkohleanalytische Untersuchungen zur Waldgeschichte der Vogesen im Tal von Miellin, *Freiburger Universitätsblätter*, 160 (2): 111-118.

NÖLKEN W., 2004.- Holzkohleanalytische Untersuchungen an Meilerplätzen in den Suedvogesen, *Zeitschrift für Archäologie des Mittelalters*, 31 (2003): 196.

NÖLKEN W., 2005.- *Holzkohleanalytische Untersuchungen zur Waldgeschichte der Vogesen.* - PhD Thesis. Universität Freiburg, 182 p.

SCHLAEPFER P., BROWN R., 1948.- *Über die Struktur der Holzkohle,* Zürich, EMPA-Bericht, 121 p. (Eidg. Materialprüfungsanstalt f. Industrie, Bauwesen u. Gewerbe, 153).

SCHWEINGRUBER F.H., 1990.- *Microscopic wood anatomy. Structural variability of stems and twigs in recent and subfossil woods from Central Europe,* Birmensdorf, Swiss Federal Institute for Forest, Snow and Landscape Research, 226 p.

# ARCHAEOLOGICAL EXPERIMENTS IN FIRE-SETTING:

# PROTOCOL, FUEL AND ANTHRACOLOGICAL APPROACH

Vanessa PY

MMSH, Laboratoire d'Archéologie Médiévale Méditerranéenne, UMR 6572 CNRS
Université de Provence - Aix-Marseille I
5 rue du Château de l'Horloge B.P. 647 F-13094 Aix-en-Provence cedex 2 (France)
py@mmsh.univ-aix.fr

Bruno ANCEL

CCSTI du Château St-Jean
Hôtel de Ville F-05120 L'Argentière-La-Bessée
brancel@wanadoo.fr

ABSTRACT: From 1997 fire-setting experiments have been undertaken each winter in the Fournel silver mines at L'Argentière-la-Bessée (Hautes-Alpes, France). The objective is to work through this "process" to rediscover technical know-how, evaluate the combined role of the fire's intrinsic (fuel) and exterior (ventilation, pyre architecture) factors, and potential results. Beyond the strict environmental aspects, this process highlights methods of fuel management, fire practice and know-how, from the forest to the mine. The pyres are set against a hard native quartzite wall, in order to pierce a gallery 1,20 m high and 1 m wide, following the observed medieval network. As the archaeological-anthracological analyses suggest, the available timber species are *Pinus* type *P. sylvestris*, *Larix-Picea* and to a lesser degree *Abies* sp. The principal parameters are systematically measured and observed : timber weight, hygrometry, size of logs, pyre makeup (set against the quartz wall, in tower form, laid horizontally), fire dynamics and surrounding temperatures. The work face advance is evaluated and measured. Rock and charcoal residues are measured granulometrically and observed macro- and microscopically. The anthracology constitutes a catalogue of anatomic deformations in this specific context. The question is to define the variable(s) producing stigmata in order to open up study perspectives on the fuel operating chain. This communication examines the preliminary studies carried out during a master degree thesis and during scheduled archaeological excavations.
KEY WORDS: mining, fire-setting, techniques, fuel management, charcoal, anatomical signatures

RÉSUMÉ: Depuis 1997 des expérimentations de taille au feu sont menées chaque hiver dans les anciennes mines de plomb argentifère du Fournel à L'Argentière-La-Bessée (Hautes-Alpes, France). L'objectif est de pratiquer cet outil pour retrouver des gestes techniques, évaluer le rôle combiné des facteurs intrinsèques au feu (combustible) et extrinsèques (ventilation, architecture du bûcher) et les rendements potentiels. Au-delà d'un aspect strictement environnemental, cette entreprise tend à caractériser des modes de gestion du combustible, des pratiques et un savoir-faire du feu, de la forêt à la mine. Les attaques sont menées contre une paroi de quartzites non altérées et très dures pour percer une galerie en travers-banc, haute de 1,20 m. et large de 1 m., à l'image des portions observées dans le réseau médiéval. Comme le suggèrent les analyses anthracologiques archéologiques, les essences sollicitées (disponibles) sont *Pinus* type *P. sylvestris*, *Larix-Picea* et dans une moindre mesure *Abies* sp. Les principaux paramètres font l'objet de mesures et d'observations systématiques: poids du bois, hygrométrie, gabarit des bûches, confection du bûcher (adossé contre la paroi, en tour, couché), dynamique du feu et températures ambiantes. L'avancement du front de taille est évalué et mesuré. Les résidus de roche et les charbons font l'objet de mesures granulométriques et d'observations macro- et microscopiques. L'anthracologie permet de constituer un catalogue de déformations anatomiques dans ce contexte spécifique. Il s'agit de définir la ou les variables à l'origine des stigmates, pour ouvrir des perspectives d'études sur la chaîne opératoire du combustible. Cet article fait état des recherches préliminaires menées dans le cadre d'un Diplôme d'Etudes Approfondies et d'une fouille archéologique programmée.
MOTS CLÉ : exploitation minière, abattage au feu, pratiques, gestion du combustible, charbons de bois, signatures anatomiques

ZUSAMMENFASSUNG: Seit 1997 werden jährlich im Winter Versuche zum Abbau erzführenden Gesteins durch Feuersetzen in den Silberminen des Fournel in L'Argentière-La-Bessée (Hautes-Alpes, France) durchgeführt. Ziel ist, die Techniken dieses Verfahrens zu rekonstruieren und hiermit die jeweilige Wirkung stoffeigener Faktoren (Brennstoff) und äußerer Einflüsse (Belüftung, Schichtung des Brennholzes) sowie die Abbauleistung einzuschätzen. Über die Frage der Umwelt hinaus befaßt sich die Studie mit der Organisation der Verwertung des Brennmaterials, den Praktiken sowie den Kenntnissen des Feuersetzens, der Holzwirtschaft und des Bergbaus. Der Abbau des unverwitterten, sehr harten Quarzitgesteins beginnt an der Felswand und treibt in diese einen 1,20 m hohen und 1 m breiten Stollen vor, nach dem Vorbild der vor Ort vorgefundenen mittelalterlichen Systeme. Nach Maßgabe der anthrakologischen

Untersuchungen wurden hierzu *Pinus* vom Typ *P. sylvestris*, vgl. *Larix-Picea* und in geringerem Umfang *Abies* sp. verwendet. Hierzu werden die Hauptdaten gemessen und systematischen Beobachtungen unterzogen: Gewicht der Hölzer, Feuchtigkeit, Masse und Form des Scheiterhaufens (Schichtung gegen die Wand, Turmform, Legung), Feuer- und Temperaturentwicklung. Der Fortschritt der Abbaukante wird bewertet und vermessen, die Abplatzungen und Holzkohlereste gemessen und makro- wie mikroskopisch untersucht. Als Ergebnis der anthrakologische Studie liegt ein Katalog der anatomischen Verformungen in diesem besonderen Kontext vor, der Einblick in die unterschiedlichen auf das Gestein einwirkenden Mechanismen gewährt und somit Ausblicke auf die Waldwirtschaft und Holzgewinnung erlaubt. Unser Beitrag stellt die Ergebnisse von Voruntersuchungen vor, die im Rahmen eines Dissertationsvorhabens durchgeführt wurden.

STICHWORTE: Bergbau, Feuersetzen, Techniken, Praktiken, Holzwirtschaft, Brennholz, Brennstoff, Holzkohle, holzanatomisch

## INTRODUCTION

When the wall-rock was particularly hard, fire-setting was the most widespread technique in the Middle Ages to extract the ore. Its success in practice does not show through in the written sources and the medieval iconography which are too lacunary and allusive. Only archaeology can bring new light, thanks to the architectural analysis of the works and examining the sterile granulometry and sedimentology. One of the most pre-eminent characteristics is the abundance of charcoal deposits, relics of the thousands of blazes which made it possible to open up and work the mine. These deposits constitute a mass of hard to interpret significant information because the perception of timber use through the human prism generates deformations. On the basis of this postulate, recourse to archaeological experimentation is inevitable. It should improve comprehension of the ground, clarify the interpretation of operational dynamics and make it possible to characterise modes of management of the deads and fuel. These prospects open possibilities of paleo-ecologic, technological and practical interpretations.

Except for laboratory tests realised during the 19th and at the beginning of the 20th c. (DAUBRÉE 1861, HOLMAN 1927), experimental fire-setting was initiated by Welsh archaeologists to study the technical incidences of tools made from stag antlers and stone on rock weakened by fire, in proto-historic mines (CREW 1990, LEWIS 1990, TIMBERLAKE 1990b). At the end of the 19th c., in the nineties, pioneer tests in France were carried out in the mines of Goutil-L'Argentario and at Melle. They showed the real potentialities of experimentation to study the interrelationships of the mine and the forest (DUBOIS 1996, TEREYGEOL 1998).

At L'Argentière-La-Bessée, 66 experiments were led during the 2002-2004 campaigns whilst profiting from a pluri-disciplinary co-operation on the interface of human action on the environment. They constitute the first steps of an in-depth study on wood and its specific uses for the mining activity in the Fournel valley. The experimental site is located quite far underground, in the engine chamber, chosen for its accessibility and the possibilities of ventilation. The transport of various materials, wood loads and tools is facilitated by roomy access, characteristic of the modern works. A comfortable workspace facilitates scientific follow-up of pyre combustion and data recording, by eliminating the risks caused by choking smoke (FIG. 1).

The objective of this contribution is not to develop an exhaustive presentation of the fire-setting experimental work in the Fournel for several years now, but to clarify the specific study of fuel and the charcoal residues to open a reflection on the mining uses of fire and the economy of timber. This original approach lies in the broader framework of current research on the history of technical know-how and types of mediaeval resource management.

**Figure 1.** Experimental operating area (modern workings, Fournel mines).

## SCIENTIFIC FOLLOW-UP OF EXPERIMENTAL SIZE FIRE-SETTING

### Periodicity of the experiments

To profit from good ventilation, experiments are held during the winter. Based on the initial work done since 1997, the rate of one fire per day was selected to allow the cooling of the rock face between each fire, and better recording of the data (collecting of residues, sifting, sorting, weighing). Indeed, at the time of the first series of experiments, the fires were launched continually for 33 hours so as not to let the rock face cool down (ANCEL AND MARCONNET 1997). In the same manner, the second series of fires in 1998 was led uninterruptedly for 40 hours. Finally, the 1999 experiments went on for 6 days, with 4 to 6 continuous fires per day (ANCEL 2000). These remarks pose the problem of the rate/rhythm of fire-settings in the Fournel between the 10th and 14th c. Under certain conditions, they could be carried out uninterrupted as in the mines of Kongsberg (17th-18th-19th c.) or, during a public holiday or on Saturdays or the day before to avoid accidents by suffocation as in the mines of Rammelsberg (Hartz) and in the medieval mines of Massa Marittima (Tuscany). But these isolated mentions, specific to an ore level and a technical framework, cannot be generalised for all sites and all periods (SIMONIN 1859, COLLINS 1893, BERG 1992, DUBOIS 1996). Moreover, the effectiveness of a fire burnt against a cold rock face must be tested. The strong variations in temperature on the surface of the rock can constitute an important factor (COLLINS 1893: 90-91).

Our initial objective was to dig out of a solid rock wall, a gallery of "medieval type" whose height and width would not exceed 1.5 m.

## Recording of the data

To profit from a corpus of recurring and statistically exploitable data, a record card was created and improved progressively during the experiments. It is composed of two distinct parts: during the experiment and the assessment. In general, an experiment is spread out over 24 hours. It begins with the selection of wood, stored under cover, outside the mine. Underground, the logs are split and prepared on a special surface. Each log is measured and weighed. The pyre is lit around midday. The wood is consumed on average for three-quarters to one hour. Cooling time of the hearth and the rock takes the remainder of the day and all the night. The next morning is devoted to collecting the fragmentation residues, with manual purging of the weakened rock, weighing of the residues (fragmentation and purging) and with construction of the pyre for the following fire, all carefully described with sketches. The experiments undertaken in 2002-2004 required, in all, six working weeks under ground.

The card has on the recto, a table for the course of the experimentation: statement of the ambient temperature at various stages of the experiment (arrival, lighting-up, fragmentation) and timing of the length of pyre burning. This recording makes it possible to characterise the rate/rhythm of combustion, the heat and radiation of the fire (calorific release of energy which heats the atmosphere).

This method is provisional, awaiting installation of thermal probes to measure variations in temperature of the rock face and in the hearth.

A second table is intended for recording the weight and size of the logs. Logs of different sections are split, or more rarely sawn, according to the type of pyre to raise. The kind of the wood is stipulated. The pyre has a specific insert for description of the layout of the hearth (floor), the pyre arrangement, steepness and contact with the rock face, the sides and vault of the gallery. Lastly, the bottom of the card is devoted to description of lighting-up and the progression of the pyre. This paper-work is reinforced by photography every 5 minutes.

The second part of the experiment corresponds to treatment of the residues. The verso of the card is composed of three tables reserved respectively for the charcoals, the rock fragments and the residues from manual purging.

A first stage consists in recovering the unburnt, half-burnt, or carbonised logs. These residues are weighed. On the postulate that the Old Men managed energy as well as possible, this residual fuel is re-used in further fires. In the same way, charcoal resulting from the incomplete combustion of the large logs are the subject of a first rough grading corresponding to anything more than one centimetre. These coals preserved in the dry are potentially recyclable for domestic or artisanal activities. A second grading consists in collecting the maximum of residual coals to reduce the work of sorting and fragmentation post combustion. Finally, the last coals are hand-sorted during spoil sifting. The filtered fractions are: > 10 mm, 10-5 mm, 5-2.5 mm, 2.5-1.6 mm, 1.6-1 mm, < 1 mm.

The fractions smaller than 4 mm are excluded from the total weighing of residual coals and are classed with sand and dust. These millimetre-length shards result from fragmentation post combustion of fragile coals whose oxidation was already quite advanced. The coals collected from each experimentation are the subject of sampling to be subjected to microscopic analysis. The advance of the rock face is measured in the form of a longitudinal profile.

## Description of a standard experiment: in practice

The experiment begins with provision and choice of fuel. In 1999, the wood had been stored in the mine near the experimental rock face. Output dropped gradually and became catastrophic on the sixth day. We then noted that the logs had become wetter because of their underground stay (to 99% of ambient humidity). Since then, we adopted the precaution of bringing the wood into the mine at the last moment. The logs are transported in bags and care is taken not to unnecessarily dirty or wet them.

On the spot, the logs are split according to the objectives of the experiment. In our configuration, the size of the rock face limits the quantity of wood to 70 kg. The splinters are recovered for lighting-up. The kindling burns in a few minutes and does not play a role in the attack of the rock but is determinant for fast firing of the pyre. It is thus important to prepare at least 2 to 3 kg of kindling.

The pyre should be prepared carefully as a good firing is related to the way it is made. The joining of several logs orients the flames, sometimes even making a very effective "blowtorch" effect. A log face to the wall can screen and protect the rock. Compact stacking lengthens fire time, and on the contrary a ventilated provision burns it very quickly. At the base, setting of a kind of floor allows good ventilation of the pyre, at least until its collapse. The structure must be made up so that it resists collapse longest; a mixture of thicker and smaller logs reconciles strength of the blaze and solidity of the pyre. As the cavity open to the fire has round walls, it is necessary to accommodate this in building a stable pyre; blocks of rock can be used for this purpose (they should not be counted in residue weighing!).

During the construction of the pyre, one should not lose sight of the objective of the experiment: to advance the boring of a gallery. If one "stuffs" the wood anyhow into the cavity, one will enlarge it, without hardly advancing the work face.

We chose two types of pyre construction. The "lying" pyre is composed of a floor which is based on the concave base of the rock face on which are piled up, in a more or less parallel way, the logs inclined against the face. The direction and the slope of the logs will direct the flames towards such or such part of the cavity: the face itself, the vault, the left or right-hand walls. They can also be laid out

in a cone to try to concentrate the flames towards a more precise zone. The "tower" pyre is composed of several stages of perpendicularly piled up logs (FIG. 2). The flames will thus tend to rise on all sides. If one arranges a kind of chimney in the centre of the tower, one obtains a blowtorch effect towards the vault. For these two types of pyres, one can also make a "screen", of broad and regular logs which are laid against the pyre and which prevent the flames from escaping towards the opening. They thus ensure progressive falling-in towards the rock face.

From one experiment to another, the size of the logs is never identical, nor the disposition of the fire, nor its intensity, nor the advance of the rock face, etc. Thus, by definition, each experiment is unique and their duplication is inevitably relative.

For convenience, lighting-up is done using newspaper and fruit-box slats. It is not very authentic, but there is no incidence on the objectives. On the other hand one notes that remainders of this kindling are found in the residual hearth at the end of the experiment.

If the kindling is judiciously placed, the pyre blazes up in a few minutes. The fire intensifies so much so that the flames lengthen on the vault of the cavity, then lengthen more and lick up to nearly a metre upward on the gallery wall. The fumes stagnate a little at the ceiling, then escape by a higher opening, thanks to the ascending draught which traverses the mine in winter. The fire blazes thus for 10 to 20 min, then the flames drop and stay within the cavity. It is often starting from this decrease of the flames that the phase of fragmentation begins.

The fragments of rock detach from the rock face, sometimes flat shards, sometimes blades of 10 cm surface, accompanied by dry crackings, different from those

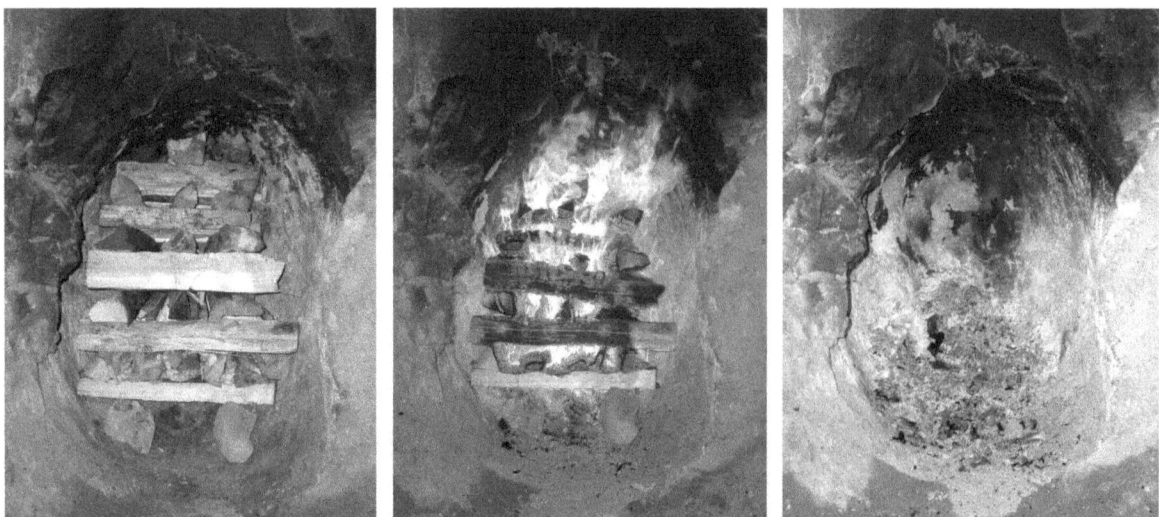

Figure 2. Experiment n°7: example of a "tower" type pyre.

| The average humidity of wood | | | | | | | | |
|---|---|---|---|---|---|---|---|---|
| Time of drying | Steres left in open air | | Steres stocked under shelter after 3 months in felling area | | Logs of 33 cm stocked under shelter 3 months after cutting | | Logs of 33 cm stocked under shelter from the shaping | |
| | Q | L | Q | L | Q | L | Q | L |
| 0 (H% init.) | 75 | 78 | 76 | 78 | 75 | 78 | 73 | 76 |
| 3 months | 48 | 62 | 48 | 61 | 44 | 61 | 36 | 40 |
| 6 months | 37 | 46 | 32 | 45 | 29 | 35 | 25 | 29 |
| 9 months | 33 | 38 | 27 | 37 | 26 | 28 | 24 | 28 |
| 1 year | 26 | 35 | 26 | 33 | 25 | 27 | 23 | 27 |
| 1 year 1/2* | 18 | 27 | 18 | 21 | 17 | 17 | 15 | 16 |
| 2 years | 16 | 24 | 16 | 17 | 16 | 14 | 14 | 13 |
| 2 years 1/2 | 15 | 24 | 15 | 18 | 15 | 14 | 13 | 13 |
| Delourme Olivier Informations from DEVILLE | | | | | | | | |

**Figure 3.** Evolution of the average humidity contained in timber after cutting.

which can occur during wood seasoning. Many of these fragmentations take place on the part of the face hidden by the pyre. On the other hand, they are quite visible from the ceiling and present a sometimes spectacular explosive character; shards of heated rock are projected up to 5 meters of distance from the hearth. The fragmentations can sometimes follow one another quickly at the time of phases of "crisis" where projections can cover thickly the front of the hearth (FIG. 2). On the sides, one observes crackings which gradually separate blades of thick rock. The fragmentations are not always spectacular, however a discrete attack operates behind and under the pyre.

After three-quarters of an hour, the burnt pyre starts to subside and tends to form an ember cluster mixed with the fragmented rock. The fragmentations are spaced in time then cease. The embers still flicker for more than an hour.

The following day, the residues of the hearth are treated. Sometimes, there remain fragments of half-burnt logs. Large charcoals mark the site of the last consumed logs and split up as soon as they are touched. During the cleaning of the hearth, one observes a great abundance of residual coals, about 10 to 20% of volume, without any comparison with what one observes in archaeological steriles. One also observes the fragmentation which happens more discreetly in the lower part of the cavity, hidden by the pyre. On the floor very thick shards are detached.

After this first cleaning, the walls are hammered to be purged. Certain zones are healthy and sound clear under the blows of the mallet, either because they were not attacked by fire, or because they were purged "naturally" during the phase of fragmentation. Other zones sound "hollow" and the masses of shards fall easily. The microfractured zones appear to resist fragmentation and must be cut

down during the purging. Dust is significant. Sifting of the product shows that close to 1/5 of the cut down rock is composed of particles smaller than 2 mm.

## ASSESSMENT AND LESSONS

Three "wood factors" can be put forward for the conduct of fire-setting: humidity content, size, the type and an "extrinsic" factor of combustion, the installation of the pyre.

## The humidity factor

The water content leads to a variation of the calorific value of the wood, with effect on the temperature from combustion and models the behavior of the fire (persistence and height of the flames). Moreover, the variation influences the duration of calcination. Good management is for example essential for fires which must "brood" a long time or blazing fires which will be consumed quickly. Experiments undertaken in 1999 show increasing fall of output which was related to fuel storage in the mine, where ambient air is saturated with moisture. This factor thus intervenes directly on rentability of fires. In accordance with the written sources and practice, it is obvious that using green heartwood is excluded for the fire. Indeed, in the Middle Ages, the conditions of lighting and ventilation could prove very difficult in major workings. This type of wood can nevertheless be used exceptionally, notably for feeding vigourous fires carried out in the open air (working starts, scrapings) for which any available timber is not negligible. The wood used for the experiences carried out in 2002 was "dry". It was delivered by a local wood merchant and stored in the form of coarsely split logs under a sheet exposed to the south. As the wood is generally quite cold because experiments are undertaken at the height of the winter, the exact water content could not measured with an electrical appliance, but it can be evaluated in a theoretical way. A log is sold dry with a water content included/understood in between 22-23 and 18%. Dried in the open air, it preserves a water content which is established between 15 and 20%. the logs used in 2003-2004 probably reached a more reduced fork with rates ranging between 15 and 18%. They could be a factor supporting the significate increase in outputs at the time of these experiences. The wood which remained is unused, these rates can be reached theoretically after a storage period under shelter or in the open air for 2 years to 2 years and half (FIG. 3). According to archaeological data, the miners of the Fournel utilised quasi exclusively coniferous trees, they thus had interest in felling in summer, when water content is the least significant. However, the traditional period of cutting of coniferous trees is at the height of winter (January-February), to avoid the abundance of resin in which they abound in spring and in summer. The obtention of a seasoned wood thus constituted a constraint, because it was necessary to be able to store it in large quantities preferably with shelter. As it is twice more profitable to dry on site the wood split in quarters rather than logs, one can consider a true operational stock preparation line of fuel upstream of the mining works. Analysis of carbonaceous residues can give elements of interpretation, thanks in particular to observation of deteriorations (fungic attacks, open slits of insect withdrawal). Elements of mycelium were observed in the cellular structures of the archaeological samples. Their study is in hand and will have to determine the theoretical time of wood stockage wood.

## The factor of size

It is possible to make blazing fires with dense wood by varying its section and water content. The more the log is thin the more the gases of combustion ignite quickly. However, to make a vigorous and durable fire, the wood surface for the flames must be significant. Fires of faggots blaze up very quickly and grow rapidly in temperature but their duration of burning is reduced. To obtain a fire long burning with a high temperature continuously, it is essential to manage the size of wood and rough-hew it according in particular to their strategic role in the fire architecture (wood of support, lighting up, reinforcements and filling). The data enregistered during the experiments carried out in 2002 (1 to 31) show that the section of the logs and the quarters constitutes a measure which intervenes in the rentability of fires. The best ratios (quantity of wood put at fire/cut down rock) correspond to the average sections of the logs ranging between 15 and 18. The lower or higher sections have not given very

profitable results. But, during the following experiments (31 to 66), more significant calibres were used, increasing significantly the average of the sections (26 to 57). Fires of significant average sections proved very profitable (FIG. 4 and 5). The management of this parameter thus plays a role more determinant than the quantity of wood sawn up. The use of large quantities of small sections is less profitable than a smaller fire, but associating large sections in combination with smaller. Inversely, to improve the output of fires of faggots, it is necessary to use a great quantity of wood, which can imply despite everything a useless enlarging of the cavity. In the case of the tower fires, the fire must flame upwards around the tower to attack the vault of the gallery. It is thus necessary to associate larger sections with smaller ones.

## The factor of species

After the criteria of size and moisture, there exists a variation of the properties of the combustion according to the species with water content varying to equal morphology. The inflammability a log is influenced by the density of the wood, its molecular composition and its content of minerals. The calorific value changes very little from one species to the other, the variations are essentially due to the chemical constitution of different species. The spruce used for the first 15 experiments is a tender light wood. Its density is low (0.45). The larch and the woodland pine used for the following experiments (in association or independently), presente a definitely higher density (0.62 and 0.55) (FIG. 4). The resin contained in these species have a very high calorific value which generate good inflammability, while the biogenic salts, also called "ashes", influence this property negatively. The more their rate is weak the more the wood will ignite quickly. This compound is present at the level of 0.3% in the woodland pine, 0.21% in the spruce and 0.17% in the larch. In the last analysis, the calorific factor varies very little between these three species. It is slightly higher in the spruce (4622 Kcal/kg) and quasi equal in the woodland pine (4556 Kcal/kg) and the larch (4597 Kcal/kg). In comparison with the experimental data, the species parameter does not directly influence the output of a fire. It appears clearly that the parameters "rate of humidity" and "size" are major. This assumption makes it possible to exclude it from a preferential selection of a species compared to another and validates the paleoecologic approache to mining charcoal. It is true that the coniferous trees are very inflammable, they are used also for the making of torches (MAGNUS 1561, CASTELLETTI AND CASTIGLIONI 1993). The use of small sizes of light deciduous trees obtains the same result. For fire setting, the peremptory necessity for a bright burning fire is not shown. The miners could use blazing fires and brooding fires according to the technical constraints to surmount, like cutting down the ceiling or the foot of a gallery. A choked fire allows transfer of heat by conduction. There is a transmission of energy of fuel towards the rock. Its propagation velocity depends on thermal conductivity over the quarzites. The heat transmitted by the flames is interesting when its radiation is channeled by the saw length. The radiation led to spectacular cracking, accompanied by projection of plates of rock heated with white. These approach the last point which constitutes in our eyes an extrinsic factor: the installation of the logs.

## The factor of fire construction

Starting from the iconographic data of the extreme end of the medieval time and modern period (general works), two types of pyres and some alternatives were used: high towers (FIG. 2) (with or without screen) or leaned (more or less vertically) against the wall of the coal face. From the strict point of view of the outputs, the first method is on average less profitable than the second (FIG. 5). From a strategic point of view, the use of these two techniques is complementary. These results are not surprising but with practice, the towers leaned and proved more direct to reach the mining objective. The tower constitutes nevertheless a good means to attack the top and to widen the cavity. Its use is effective or even more profitable to open and widen a cavity. These observations show the interest of a "slow" fire whose attacks are less spectacular but prove efficient to bore the floor and to tackle the face. Management of the orientation and the slope of the logs also makes it possible to target the attack to the left or right-hand sides of the gallery. These principal parametres are obviously not the only ones to intervene on the fire strength. The ventilation (lack of air decreases the temperature of combustion), the ambient temperatures (between the interior and the exterior) and the morphology of the rock constitute as many new factors. It seems complex to test all information recorded with classic statistical tests because the influencing parameters are too numerous. Such a step requires recourse to factorial analysis of correspondences (AFC) to determine the characters whose impact can be measured on the 66 experiments. Necessary parametres are the water content, the section of the wood, the number of logs, the species, weight and morphology of the logs, the surrounding temperature and the temperatures of combustion. This step will be possible when all the parametres can be measured. In fact, the temperatures of combustion and the gradient of rise in temperature constitute the link lacking in this experimental study. The question has to precisely determine from which temperature the rock cracks and is split. This information can clarify the problem of which fire to use: "bright burning" or "slow" fire. Does heat have to emerge gradually to do profound heating or is necessary it that it gets bright violently to cause a thermal shock ? To carry out a good attack thus requires control of the state and morphology of the wood. The constraints, techniques

and urgent economic requirements imply a rigorous management of this raw material. In this direction, mining is implicitly related to forestry development.

## ANTHRACOLOGICAL ANALYSIS OF EXPERIMENTAL CHARCOALS

Beyond taxinomic determination, anthracological study allows a specific insight on fuel and its uses. From this point of view, precursory work on charcoals resulting from paleolithic hearths and laboratory experiments to reproduce the same deformations, highlighted the interest of analysis of anatomical modifications of the structure of wood in terms of economy of timberings (THÉRY-PARISOT 1998, 2001). The archaeological charcoals found in the mines of Fournel present anatomical deformations which were observed in a recurring way on the transverse and radial levels. They are mainly radial slits often open, cellular fusions more or less partial (vitrification) and localised tangential crackings on the level of the rings of growth. Within the framework of our approach, it is possible to check if these deformations are reproductible in an experimental context. The goal is to determine the causes responsible for their appearance under the effect of carbonization in this specific context, to open prospects for studies in terms of practice. But as the principal parametres of the experimentation are not completely controlled, the possibilities of interpretation are limited. Despite everything, we considered it necessary to announce our preliminary observation to continue to feed reflexion on this subject. A sampling "test" of experimental residual coals was carried out to constitute a first inventory of recurring and significant markings. The anatomical study was carried out on 170 experimental coals resulting from 17 experiments. Measurements of the deformations (slits and cracks) and their statistical counting were not considered to be necessary because the studied corpus is incomplete. The first observations are nevertheless worthy of comments.

## The radial slits

A little more than 60% of the analyzed experimental samples present radial slits often open. One finds 65% of the individuals presenting such deformations for the experiments carried out with spruce and 66% of the individuals for those carried out with pine and/or larch. The frequency of the slits is significant with the three species. These strong percentages cannot be related to the use of a green heart or wet wood. Experiments carried out in laboratory, by controlling the parametres of carbonization perfectly, show that it is not possible to determine starting from this only criterion the use of a green heart or a seasoned wood (THÉRY-PARISOT 2001: 56-68). The appearance of shrinkage cracks is then

| n° | Logs | Type | Species of wood | Average | Sect° | Stan. deviatio | Wood stacking | Wood burned | Charcoal | ch/wood | fire-setting | Purge | Rock extrac. | Ratio |
|---|---|---|---|---|---|---|---|---|---|---|---|---|---|---|
| 1 | | A | E | | | | 31,0 | 30,8 | 0,4 | 0,013 | 5,013 | 12,279 | 17,29 | 0,56 |
| 2 | 22 | A | E | 1,1 | 15,4 | 3,1 | 23,1 | 23,0 | 0,4 | 0,018 | 5,047 | 6,823 | 11,87 | 0,52 |
| 3 | 22 | A | E | 1,3 | 17 | 3,1 | 28,4 | 28,1 | 0,4 | 0,015 | 4,03 | 9,918 | 13,95 | 0,50 |
| 4 | 28 | A | E | 1,1 | 15,1 | 3,6 | 32,1 | 32,1 | 0,4 | 0,013 | 6,294 | 7,588 | 13,88 | 0,43 |
| 5 | 27 | A | E | 1,2 | 16 | 3,2 | 31,1 | 31,1 | 0,4 | 0,014 | 7,299 | 9,431 | 16,73 | 0,54 |
| 6 | 23 | A | E | 0,9 | 16,1 | 3,9 | 21,8 | 21,7 | 0,5 | 0,022 | 5,643 | 4,578 | 10,22 | 0,47 |
| 7 | 19 | A | E | 0,9 | 16,9 | 4,4 | 17,3 | 17,0 | 0,3 | 0,018 | 4,365 | 2,935 | 7,30 | 0,43 |
| 8 | 36 | T | E | 0,8 | 18,3 | 5 | 29,4 | 28,4 | 0,4 | 0,013 | 12,924 | 8,02 | 20,94 | 0,74 |
| 9 | 25 | T | E | 0,9 | 19,5 | 5,8 | 23,5 | 23,4 | 0,2 | 0,009 | 8,1 | 7,68 | 15,78 | 0,67 |
| 10 | 31 | T | E | 0,7 | 19 | 5,7 | 22,3 | 22,3 | 0,4 | 0,020 | 8,566 | 11,902 | 20,47 | 0,92 |
| 11 | 27 | T | E | 1,0 | 17,3 | 6,1 | 26,9 | 26,9 | 0,4 | 0,014 | 7,59 | 7,47 | 15,06 | 0,56 |
| 12 | 21 | T | E | 1,1 | 17,3 | 5,3 | 22,3 | 22,0 | 0,3 | 0,013 | 4,175 | 4,638 | 8,81 | 0,40 |
| 13 | 23 | T | E | 1,0 | 17,8 | 5,2 | 24,1 | 24,0 | 0,3 | 0,012 | 8,13 | 6,73 | 14,86 | 0,62 |
| 14 | 27 | T | E | 1,2 | 18,3 | 5,6 | 32,2 | 32,2 | 0,4 | 0,013 | 7,823 | 8,157 | 15,98 | 0,50 |
| 15 | 24 | T | E | 1,2 | 17,5 | 4,5 | 28,8 | 28,6 | 0,5 | 0,017 | 9,585 | 9,68 | 19,27 | 0,67 |
| 16 | 33 | T | PM | 1,0 | 18,8 | 4,9 | 34,5 | 34,4 | 0,5 | 0,014 | 7,275 | 9,79 | 17,07 | 0,50 |
| 17 | | T | PM | | | | 30,0 | 29,5 | 0,6 | 0,020 | 6,425 | 6,425 | 12,85 | 0,44 |
| 18 | 25 | T | PM | 1,2 | 21,9 | 8,8 | 30,6 | 30,0 | 0,6 | 0,019 | 4,951 | 4,635 | 9,59 | 0,32 |
| 19 | 26 | T | PM | 1,1 | 19,1 | 7,8 | 29,6 | 29,5 | 0,7 | 0,023 | 9,13 | 12,155 | 21,29 | 0,72 |
| 20 | 28 | T | P | 1,0 | 17,6 | 4,6 | 26,7 | 26,4 | 0,6 | 0,023 | 5,299 | 3,99 | 9,29 | 0,35 |
| 21 | 29 | T | M | 1,0 | 18,4 | 6,3 | 28,9 | 28,8 | 0,9 | 0,031 | 11,934 | 5,486 | 17,42 | 0,61 |
| 22 | 23 | T | P | 1,3 | 20 | 7,2 | 29,6 | 29,4 | 0,7 | 0,023 | 8,375 | 5,984 | 14,36 | 0,49 |
| 23 | 29 | A | M | 0,9 | 18 | 5,2 | 26,1 | 25,5 | 0,6 | 0,022 | 7,555 | 5,066 | 12,62 | 0,49 |
| 24 | 17 | A | P | 1,3 | 18,3 | 4,7 | 21,7 | 21,0 | 0,5 | 0,021 | 9,372 | 5,906 | 15,28 | 0,73 |
| 25 | 22 | A | M | 1,0 | 17,9 | 4,6 | 22,6 | 21,7 | 0,6 | 0,028 | 8,005 | 6,155 | 14,16 | 0,65 |
| 26 | 20 | A | P | 1,1 | 15,6 | 4,9 | 21,1 | 20,4 | 0,5 | 0,023 | 4,12 | 3,925 | 8,05 | 0,39 |
| 27 | 20 | A | M | 1,6 | 22,7 | 9,3 | 31,3 | 31,1 | 0,7 | 0,021 | 13,055 | 6,225 | 19,28 | 0,62 |
| 28 | 19 | A | P | 1,2 | 18,4 | 8,6 | 22,6 | 22,0 | 0,7 | 0,030 | 5,36 | 4,295 | 9,66 | 0,44 |
| 29 | 34 | A+ | MEP | 1,6 | 22,5 | 7,4 | 53,9 | 52,9 | 1,1 | 0,021 | 30,315 | 17,775 | 48,09 | 0,91 |
| 30 | | A | MP | | | | 35,0 | 35,0 | 0,5 | 0,014 | 37,46 | 17,46 | 54,92 | 1,57 |
| 31 | 20 | A | M (P) | 1,7 | 26 | 10,4 | 33,8 | 33,4 | 0,5 | 0,015 | 12,1 | 4 | 16,10 | 0,48 |
| 32 | 17 | A | M (P) | 2,1 | 29 | 20 | 35,9 | 33,7 | 0,6 | 0,015 | 13,6 | 4 | 17,60 | 0,52 |
| 33 | 17 | A | M | 2,5 | 34,3 | 12,8 | 42,6 | 41,9 | 0,7 | 0,017 | 15,4 | 5,2 | 20,60 | 0,49 |
| 34 | 20 | C | M | 1,5 | 23,9 | 9,3 | 29,7 | 28,9 | 0,7 | 0,023 | 11,5 | 9,6 | 21,10 | 0,73 |
| 35 | 16 | C | M | 2,0 | 31,8 | 15,4 | 31,5 | 30,2 | 1,0 | 0,030 | 6,2 | 6,7 | 12,90 | 0,43 |
| 36 | 21 | C | M | 1,5 | 25,2 | 13,7 | 31,7 | 31,1 | 0,6 | 0,017 | 6 | 6,5 | 12,50 | 0,40 |
| 37 | 25 | T | M | 2,2 | 31,9 | 19,2 | 53,9 | 53,6 | 0,7 | 0,013 | 15,7 | 18 | 33,70 | 0,63 |
| 38 | 19 | T | M | 3,1 | 43,6 | 14,9 | 59,1 | 58,7 | 1,0 | 0,017 | 21,3 | 13,2 | 34,50 | 0,59 |
| 39 | 20 | T | M | 2,7 | 41,3 | 21,8 | 53,4 | 53,3 | 1,0 | 0,018 | 13 | 15 | 28,00 | 0,53 |
| 40 | 14 | A | M | 3,0 | 40,5 | 26,9 | 42,5 | 40,1 | 0,4 | 0,010 | 10,6 | 4,9 | 15,50 | 0,39 |
| 41 | 11 | A | M | 3,6 | 43,6 | 19,3 | 39,4 | 38,3 | 0,8 | 0,019 | 19,2 | 10,25 | 29,45 | 0,77 |
| 42 | 16 | A | M | 2,5 | 36,1 | 17,2 | 40,6 | 39,6 | 0,3 | 0,009 | 19,4 | 9,1 | 28,50 | 0,72 |
| 43 | 27 | TE | M | 2,2 | 36,1 | 14,6 | 60,3 | 60,3 | 0,9 | 0,015 | 24,6 | 21 | 45,60 | 0,76 |
| 44 | 30 | TE | M | 2,2 | 41,8 | 20,4 | 65,3 | 64,0 | 0,6 | 0,009 | 21,9 | 15 | 36,90 | 0,58 |
| 45 | 28 | TE | M | 2,0 | 42 | 19,7 | 56,9 | 55,8 | 1,2 | 0,020 | 12,8 | 9,3 | 22,10 | 0,40 |
| 46 | 25 | TE | M | 2,2 | 39,1 | 14,1 | 55,1 | 24,9 | 0,9 | 0,017 | 11,3 | 9,25 | 20,55 | 0,83 |
| 47 | 23 | TE | M | 2,6 | 39,5 | 20,1 | 58,9 | 58,4 | 0,7 | 0,011 | 19,9 | 19 | 38,90 | 0,67 |
| 48 | 19 | A | M | 2,0 | 30,3 | 18,7 | 38,6 | 37,9 | 0,7 | 0,018 | 15,9 | 4,3 | 20,20 | 0,53 |
| 49 | 20 | A | M | 2,3 | 33,4 | 21,1 | 46,5 | 46,4 | 1,3 | 0,029 | 14,23 | 5,14 | 19,37 | 0,42 |
| 50 | 19 | A | M (P) | 2,2 | 30,5 | 14,4 | 40,9 | 40,8 | 0,9 | 0,023 | 15,765 | 7,805 | 23,57 | 0,58 |
| 51 | 18 | A | M | 2,9 | 38,5 | 16,7 | 53,0 | 53,0 | 1,3 | 0,025 | 24,009 | 13,355 | 37,36 | 0,70 |
| 52 | 27 | A | M | 1,7 | 24 | 9,5 | 46,0 | 45,6 | 1,2 | 0,025 | 9,815 | 10,185 | 20,00 | 0,44 |
| 53 | 15 | AC | M | 3,3 | 51,5 | 29,9 | 48,9 | 48,7 | 1,2 | 0,024 | 26,175 | 8,97 | 35,15 | 0,72 |
| 54 | 17 | CT | M | 2,8 | 37,6 | 26,8 | 47,7 | 47,2 | 0,7 | 0,014 | 24,44 | 14,655 | 39,10 | 0,83 |
| 55 | 13 | C | M | 3,3 | 40,1 | 19,7 | 43,5 | 43,2 | 0,7 | 0,016 | 11,6 | 9,76 | 21,36 | 0,49 |
| 56 | 12 | C | M | 3,1 | 37 | 21,9 | 37,1 | 36,6 | 0,6 | 0,016 | 9,96 | 5,535 | 15,50 | 0,42 |
| 57 | 12 | C | M | 3,3 | 44,2 | 25,2 | 40,1 | 37,5 | 0,3 | 0,007 | 12,23 | 6,2 | 18,43 | 0,49 |
| 58 | 12 | A | M | 4,0 | 49,6 | 30,5 | 48,0 | 47,9 | 1,1 | 0,024 | 17,99 | 9,1 | 27,09 | 0,57 |
| 59 | 12 | A | M | 3,2 | 36,7 | 23,1 | 38,6 | 37,8 | 0,8 | 0,020 | 16,207 | 9,708 | 25,92 | 0,69 |
| 60 | 20 | A | M | 3,3 | 41,5 | 26 | 66,7 | 65,2 | 1,2 | 0,017 | 61,513 | 24,63 | 86,14 | 1,32 |
| 61 | 10 | A | M | 3,0 | 43,7 | 27,7 | 30,0 | 28,4 | 0,7 | 0,023 | 4,965 | 6,92 | 11,89 | 0,42 |
| 62 | 13 | A | M | 5,3 | 67,6 | 26,5 | 69,3 | 68,8 | 2,2 | 0,031 | 57,12 | 26,34 | 83,46 | 1,21 |
| 63 | 15 | A | M | 4,0 | 46,3 | 25,9 | 59,7 | 59,7 | 1,7 | 0,028 | 35,04 | 25,78 | 60,82 | 1,02 |
| 64 | 13 | A | P | 3,6 | 47,2 | 14,5 | 46,9 | 46,9 | 0,3 | 0,006 | 19,59 | 16,815 | 36,41 | 0,78 |
| 65 | 17 | AC | M | 4,2 | 52,7 | 23,3 | 71,4 | 71,4 | 1,7 | 0,024 | 54,8 | 19,9 | 74,70 | 1,05 |
| 66 | 13 | A | M | 4,6 | 57,4 | 19,3 | 60,3 | 59,7 | 1,1 | 0,018 | 23,5 | 13,9 | 37,40 | 0,63 |

Ratio > 1
Ratio > 70 < 1
Ratio > 50 < 70
Ratio < 50

E = *Picea abies*
M = *Larix europaea*
P = *Pinus sylvestris*

A = standing wood
T = tower
C = lain wood
TE = tower + screen

AC = standing wood-lain wood
CT = lain wood-tower

**Figure 4.** General table of measurements taken during the fire-setting experiments.

interpreted as the effect of several factors combined, different from one species with another. Indeed, the behaviour of wood on fire depends, upstream, of its particular characteristics and, downstream, of the whole of variables of carbonization. Our results tend to show for the three species used, a regularity of the occurrence of splits. Consequently, the factors of carbonization seem significant. To validate this assumption, it is necessary to record the temperatures in contact with the rock and in the hearth, as well as the gradient of rise in temperature. The

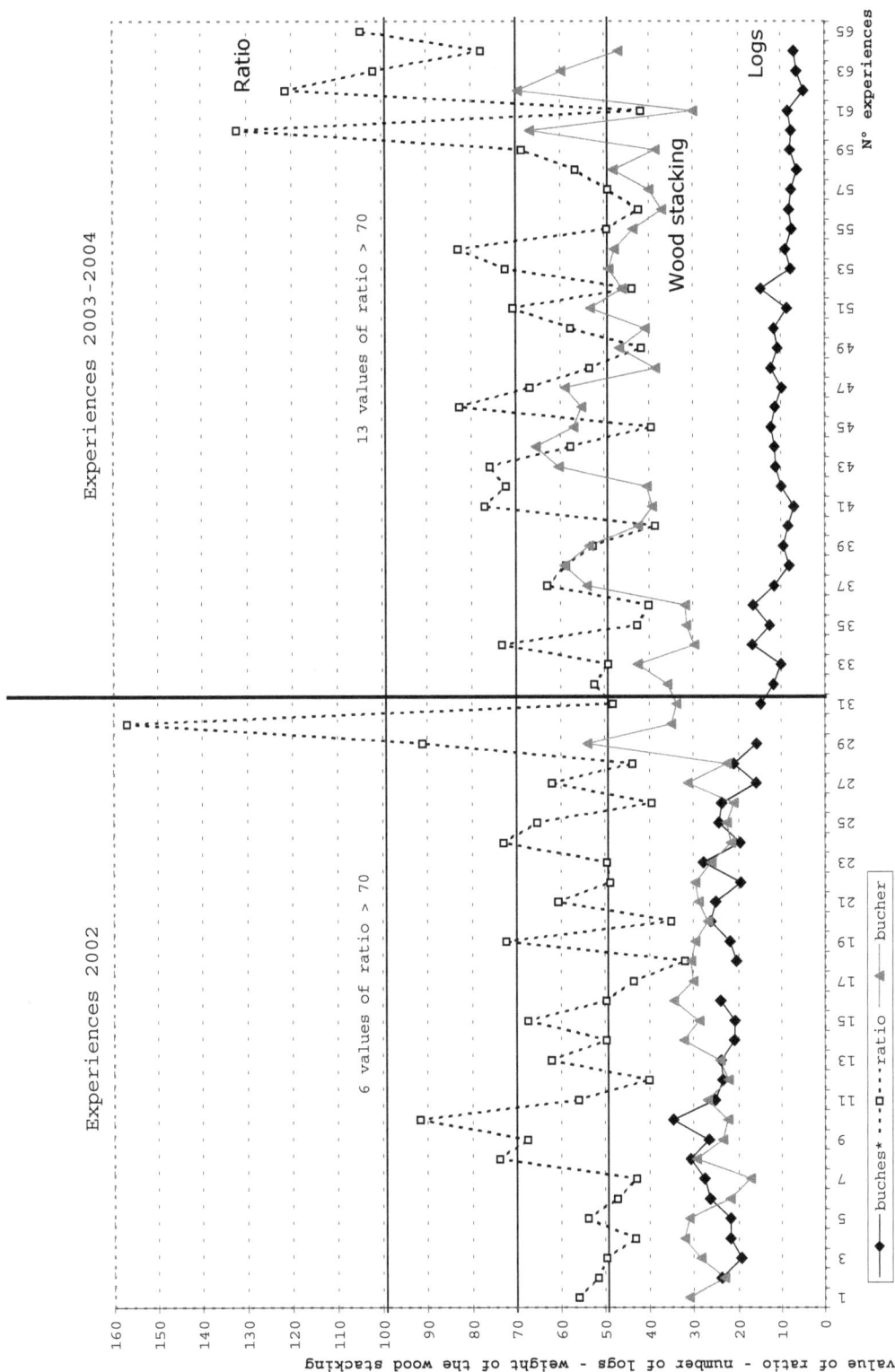

**Figure 5.** Comparative chart of pyre weight, number of logs and ratio for the 66 experiments.

observations carried out by other authors show that the temperatures can easily go up up to 800 and 900 degrees in context of fire setting. The experiments undertaken in the laboratory by I. Théry-Parisot were limited to 750 degrees. It is thus not excluded that higher temperatures associated with a confined atmosphere are at the origin of the phenomenon. These remarks propose the need for parallels to studying the appearance of the deformations within the framework of archaeological experimental analyses carried out in laboratory.

## Vitrification

Vitrification is the second deformation observed in a recurring way in the experimental samples. This phenomenon regularly announced by anthracologists is often associated with artisanal contexts like residual deposits of charcoals or the hearths of furnaces of potters (reducing cooking). Some see the result of a thermal shock there (water sprinkling for example), others determine the marks of the carbonization of a green heart or wet wood. To include/understand the origin of this phenomenon thus constitutes a key of interpretation to characterise stages of the operational chain of fuel and know-how of fire. This deterioration is caused by the fusion and homogenisation of the anatomical structures of wood which lead sometimes to the disappearance of certain criteria of determination. The seen samples have a vitreous aspect, shining and a globular structure very characteristic, but this level of degradation remains specific. Analyzed experimental charcoals contain a high percentage of partially vitrified fragments. Indeed, more or less advanced stages were distinguished. For a portion of them, the woody cells present a slightly vitreous aspect, perceptible out of radial cut. Others, more difficult to break manually, present a total fusion of the anatomical elements producing a slag aspect which seldom reaches all the mass of the sample. The most deteriorated samples come from wood which had evacuated a fluid substance, at the time of the phases of drying and dehydration, become blackish with the outburst of the carbon dioxide. It is probably to do with the combustion of the resin which forms sometimes significant pockets in woody fibres. The coniferous trees are gorged with it in spring and in summer. Raised at high temperature, they could have an impact on the state of the charcoals. In this case, if these deformations are generated by abundant resin, their determination could offer indications over the season of cutting. It thus releases a true archaeological potential from this type of approach in particular with regard to the "signatures" of the practices of supply wood and cutting to fire. The experiments show very clearly that the specific methods of this technique generate notorious phenomena on the level of the anatomy of wood. However, the control of all the parametres which influence the behaviour of wood to fire and the output requires a considerable material investment (thermal probes and analyser of images).

## Prospects: crossed glances with archaeology

In continuity with this study, is the question of weaving bonds between experimentation, groundwork and anthracologic analysis. The mine constitutes an original archaeological framework which imposes adapted methods. It is in particular a question of specifying potential production in charcoal residues and of evaluating their level of representativity in terms of provision territory and practices. A recording of the quantities produced for each fire was carried out for grainage higher than 10 mm, 10-5 mm and lower than 5 mm. It appears very clearly that pine and larch produce more residues than spruce. The quantity of charcoals obtained does not depend on the quantity of wood put on the pyre but on the type. One should not however not lose sight of the fact that the humidity factor which can exploit a considerable role the reduction of mass and the number of fragments (LOREAU 1994). A test of $X^2$ shows that the distribution of the characters "quantity of wood" and "production of charcoal" differs in a highly significant way with a probability higher than 0.001. In the same way, the fine fractions increase to a significant degree with the use of pine and larch. There is thus a differential production of residues which depends partly on the species. The conditions of combustion influence in a more or less implicit way the production of charcoals in particular in the case of "brooding" fire. The question of more precisely defining this remark will depend on factorial analysis of the correspondences.

## CONCLUSION

This contribution establishes the bases of an experimental protocol of the size to fire and establishes potential work directions. Indeed, this technique proves much more complex than the iconography and the texts make it appear. Access to know-how and practice by experimentation awakens a true "knowledge of burning" and "knowledege of managment" of fuel which today is completely "diluted" and skewed by uses of fire for pleasure and authenticity. Admittedly archaeological experimentation constitutes a working tool which is still imperfectly controlled and always used advisedly. Nevertheless, repetition of the protocol, scientific follow-up of fires and multiplication of tests to solve the various technical difficulties, make it possible to avoid the subjective character which is often charged to it. Gestures and skills are acquired gradually and highlight a long operational chain which starts with the provision of fuel and its storage, then continues with choice of wood and its preparation, bringing into the mine, construction of the pyre and its lighting, and finishes with treatment of the products, purging, sorting and storage of residues. The minerallurgic operational chain then takes over. Analysis of the products of experimental fragmentation, compared with that of the archaeological steriles in particular, highlights the complexity of underground procedure (ANCEL ET AL, IN PRESS). The outputs obtained, probably quite poor taking into consideration those which were reached by experienced minors, are useful in the reflection over working time, considerable for the comprehension of operational dynamics and exploitation strategies, and also for the potential impact on forest dynamics.

Wood was a paramount energy source, of which it is necessary to understand the context. Perfect knowledge of its properties and management, adaptation to environmental possibilities and technical constraints, enable it to be used in an effective and very profitable way. The experiments highlight factors independent of the species validating a paleo-ecologic approach to the residues of mining activity. It helps to reconstruct the evolution of a medieval industrial landscape. Concerning forests, archaeological charcoals truly constitute the only source of information. The experimentation clarifies our perception of the landscape and allow reflection on the relevance of the archaeological charcoal deposits and their representativity in terms of territory of provision. Moreover, it stresses practices both of the operational fuel chain, thanks to analysis of the phenomena of combustion, and of the origin of anatomical deformation and deterioration. This research orientation, too often neglected, has the merit of offering great prospects to archaeological anthracology.

ACKNOWLEDGMENTS

The authors make a point of addressing very particular thanks to Mr. Ian Cowburn who took care of the tiresome translation work, and Mr. Andreas Hartmann-Virnich for the translation of the German summary.

This study continues within the framework of programmed archaeological excavations directed by Mr. Bruno Ancel and is part of a pluri-disciplinary co-operation organised around the Joint Action: *To know how to burn, to know how to manage fuel supply in the southern potting and mining areas (11th-16th c.)*, co-ordinated by Mrs. Aline Durand.

REFERENCES

ANCEL B., 2000.- *Les mines d'argent du Fournel, L'Argentière-La-Bessée, Hautes-Alpes, Région PACA, Fouilles programmées 1998-2000*, Document Final de Synthèse, L'Argentière-la-Bessée, C.C.S.T.I.: 59-60.

ANCEL B., MARCONNET C., 1997.- *Les mines d'argent du Fournel, L'Argentière-La-Bessée, Hautes-Alpes, Région PACA, Fouilles programmées 1995-1997*, Document final de synthèse, L'Argentière-la-Bessée, C.C.S.T.I: 44-50.

ANCEL B., PY V., MARCONNET C., IN PRESS.- De l'usage minier du feu: à l'interface homme et environnement. Sources et expérimentations, *Cahier d'Histoire des Techniques*, Publications de l'Université de Provence.

BERG B.I., 1992.- Les techniques d'abattage à Kongsberg (Norvège) du XVIIe au XIXe siècle: pointerolle, travail au feu et tir à la poudre, *in: Les techniques minières de l'Antiquité au XVIIIe siècle. Actes du colloque international sur les ressources minières et l'histoire de leur exploitation de l'Antiquité à la fin du XVIIIe siècle (Strasbourg, avril 1988)*, Paris, Editions du C.T.H.S.: 55-76.

CASTELLETTI L., CASTIGLIONE E., 1993.- Resti lignei del XII-XIII secolo dalla miniera "VIII Sfera", *in: Milano e la Lombardia in éta comunale secoli XI-XIII (Milano-Palazzo Reale 15 aprile)*, Silvana, 11 luglio 199: 239-242.

COLLINS A. L., 1893.- Fire-sitting: the art of mining by fire, *Journal of Federate Institute of mining engineers*, V: 82-92.

CREW P., 1990.- Firesitting experiment at Rhiw Goch, 1989, *in: P. Crew and S. Crew (eds), Early Mining in the British Isles. Proceedings of the Early Mining Workshop (Plas Tan y Bwlch Snowdonia, 17-19 November 1989)*, Plas Tan y Bwlch Occasional Paper, 1: 57.

DAUBRÉE A., 1861.- Emploi de la chaleur et de la décrépitation qu'elle peut produire pour le percement de certaines roches très dures, et notamment des quartzites, *Annales des mines*, 5ème série, XIX, Mémoires: 23-25.

DUBOIS C., 1996.- L'ouverture par le feu dans les mines: histoire, archéologie et expérimentation, *Revue d'Archéométrie*, 20: 33-46.

HOLMAN B. W., 1927.- Heat-treatment as a Agent in Rock-breaking, *Transactions of the Institute of Mining Metals*, 36: 219-262.

LEWIS A., 1990.- Fire-setting experiments on the Great Orme, 1989, *in: P. Crew and S. Crew (eds), Early Mining in the British Isles. Proceedings of the Early Mining Workshop (Plas Tan y Bwlch Snowdonia, 17-19 November 1989)*, Plas Tan y Bwlch Occasional Paper, 1: 55-56.

LOREAU P., 1994.- *Du bois au charbon de bois: approche expérimentale de la combustion*, DEA, Université de Montpellier II, 64 p.

MAGNUS OLAUS, 1561.- *Historia de gentibus septentrionalibus...* Antwerp, Editions Christophe Plantin: 28.

SIMONIN L., 1859.- De l'ancienne loi des mines de la République italienne de Massa-Marittima (Toscane), *Annales des Mines*, 5ème série, XV, Partie administrative: 1-15.

TEREYGEOL F., 1998.- Les mines de Melle (Deux-Sèvres): Une expérimentation d'attaque au feu *in situ, in: L'innovation technique au Moyen Age. Actes du VIe congrès international d'archéologie médiévale (Dijon, Mont-Beuvray, Chenôve, Le Creusot, Montbard, 1-5 octobre 1996)*, Paris, Editions Errance: 111-113.

THÉRY-PARISOT I., 1998.- *Economie du combustible et paléoécologie en contexte glaciaire et périglaciaire, Paléolithique moyen et supérieur du Sud de la France. Anthracologie, expérimentation et taphonomie*, Doctorat, Université de Paris I, 499 p.

THÉRY-PARISOT I., 2001.- *Economie des combustibles au Paléolithique. Expérimentation, taphonomie, anthracologie*, Paris, Editions C.N.R.S. - C.EP.A.M., 195 p. (*Dossier de Documentation Archéologique*, 20)

TIMBERLAKE S., 1990a.- Review of the historical evidence for the use of firesetting, *in:* P. Crew and S. Crew (eds), *Early Mining in the British Isles. Proceedings of the Early Mining Workshop (Plas Tan y Bwlch Snowdonia, 17-19 November 1989)*, Plas Tan y Bwlch Occasional Paper, 1: 49-52.

TIMBERLAKE S., 1990b.- Firesetting and primitive mining experiment, Cwmystwyth, 1989, *in:* P. Crew and S. Crew (eds), *Early Mining in the British Isles. Proceedings of the Early Mining Workshop (Plas Tan y Bwlch Snowdonia, 17-19 November 1989)*, Plas Tan y Bwlch Occasional Paper, 1: 53-54.

WILLIES L., 1994.- Firesetting technology, *Bulletin of the Peak District Mines Historical Society*, 12 (3): 1-8.

# TRES MONTES (NAVARRA, SPAIN): DENDROLOGY AND WOOD USES IN AN ARID ENVIRONMENT

Yolanda CARRIÓN MARCO

Departamento de Prehistoria y Arqueología, Universitat de València
Avda. Blasco Ibáñez, E-28 46010 Valencia (Spain)
Yolanda.carrion@uv.es

**ABSTRACT:** In this paper, the results of the dendrological analysis of carbonized material from Tres Montes dolmen (Bardenas Reales, Navarra) are offered. This material comes from the wooden structure of the monument. For this task, it has been used a single taxon, *Juniperus* sp. The aim of this analysis is to achieve some information about the development of plant communities in the surrounding areas of the site, the existing vegetal resources and human exploitation patterns. Dendrological analyses are perfectly adapted to this research, since ecological and climatic questions, and human evidences can be inferred from tree growth ring observation and measurement.
**KEY WORDS:** Bardenas Reales, *Juniperus,* Dendrology, Growth Rings

**RÉSUMÉ:** Dans cet article, nous présentons les résultats de l'analyse dendrologique des charbons de bois du dolmen de Tres Montes (Bardenas Reales, Navarra). Tout le matériel a été prélevé de la structure en bois du monument. Pour ce travail, un seul taxon a été utilisé, *Juniperus* sp. L'analyse dendrologique peut être une importante source de données sur l'environnement végétal proche du lieu d'habitat, les interactions des hommes avec le paysage ou l'exploitation humaine des ressources végétaux, puisque la largeur et morphologie des cernes de croissance peuvent enregistrer les principaux événements climatiques et anthropiques.
**MOTS CLÉS:** Bardenas Reales, *Juniperus,* dendrologie, cernes de croissance

**ZUSAMMENFASSUNG:** Dieser Bericht zeigt das Ergebnis der dendrologischen Untersuchung aus dem Kohlenstoff von dolmen Tres Montes (Bardenas Reales, Navarra). Dieser Stoff stammt aus der Struktur des Holzbaus diese Denkmales. Für deren Bau wurde nur das taxon verwendet: *Juniperus* sp. Das Ziel dieser Forschung war Informationen über die Pflanzengesellschaft in diesem Ort zu erreichen, die menschliche Ausnutzung von Planzenmitteln. Die dendrologische Untersuchung passt sich vollkommen an diese Forschung an, da über die Jahresringe der Pflanzen Informationen über die klimatischen, ökologischen und Völkerbeschreibung zu finden sind.
**STICHWORTE:** Bardenas Reales, *Juniperus*, Dendrologie, jahrringe

## INTRODUCTION

The Bell-Baker dolmen of Tres Montes is located next to the Ebro Valley, 370 m.o.s.l., in a natural region called Bardenas Reales (FIG. 1). The current landscape of the region depends on its geographical situation, since it is located between two important mountain-ranges, which don't allow humid winds to get this area. Because of the continental influence, the annual rains hardly get 400 mm, as torrential rainfalls, with a long summer droughts and very cold winters. Temperatures can fluctuate throughout the year from 40°C in the summer, to -5°C in the winter. For all these reasons, erosive processes are very intense

and there is a characteristic relief with high plains and deep gorges, and vegetation is subdesertic and steppe-like (ELÓSEGUI ALDASORO AND URSÚA SESMA 1994, ELÓSEGUI ALDASORO ET AL. 1990).

Erosion and torrential rains don't help vegetation development, and it rarely appears all around the site, although there are some formations which perfectly adapt to these biogeographical conditions. Tres Montes is located on a little top-planed hill, with shrubs of *Rosmarinus officinalis* and several species of *Thymus* developing on the slopes. In fact, this is the most spreaded vegetation in the region, especially on rocky slopes and

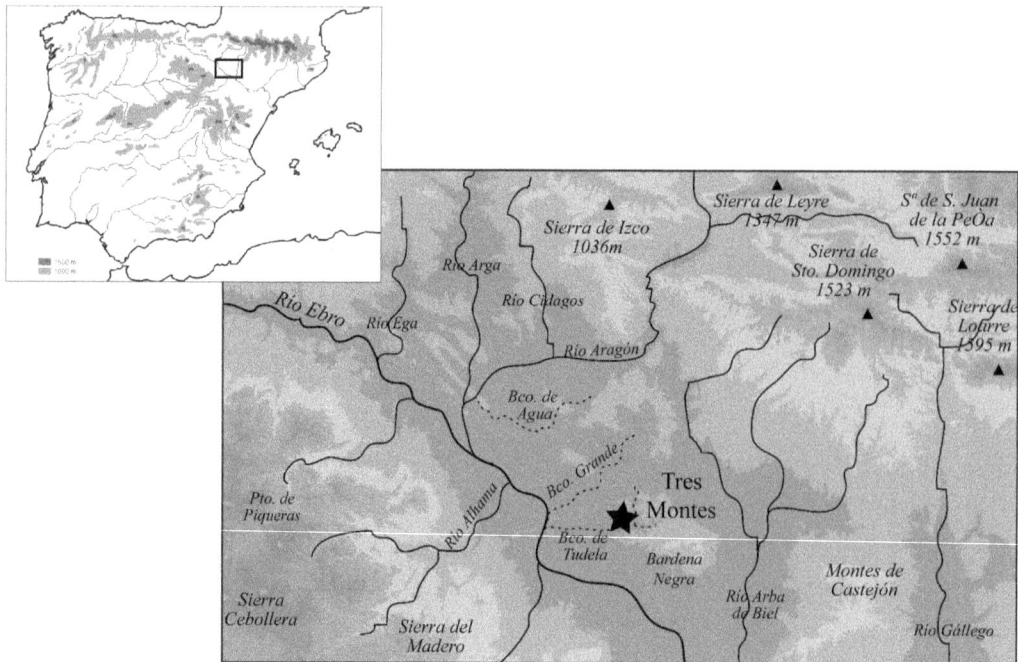

**Figure 1.** Situation map of Tres Montes site.

**Figure 2.** Current landscape in the near environment of Tres Montes before fielwork in the site (picture by J. Sesma Sesma).

poorly developed soils, but recently expanding to other areas because of the human tree felling and animal grazing (FIG. 2). Other species are present, such as *Linum suffruticosum*, some legumes, several species of Cistaceae (*Helianthemum cinereum subsp. rubellum, Helianthemum pillosum, Fumana ericoides, Fumana thymiflora*), and *Buxus sempervirens* in northern slopes (ELÓSEGUI ALDASORO AND URSÚA SESMA 1994).

There are some *Tamarix canariensis* formations along the streams running NE-SW. They are adapted to the salty soils and sandy sediments carried by the torrential waters, together with *Juncus* and Gramineae; where soils are too salty, vegetation is only composed by Chenopodiaceae and *Stipa tenacissima*.

There is some tree vegetation on slopes with higher

rains (over 500 mm), with *Pinus halepensis* and a well-developed bushy layer of *Quercus coccifera, Rhamnus lycioides, Phillyrea angustifolia, Rhamnus alaternus, Juniperus oxycedrus, Juniperus phoenicea, Juniperus thurifera* and *Pistacia lentiscus*. Only in farther north slopes with deeper soils, some *Quercus rotundifolia* and *Arbutus unedo* develop. Human exploitation of plant resources from prehistoric times does not allow valuing the real composition of natural vegetation, but this must have always been limited by the climatic characteristics of the region.

## MATERIAL AND METHODS

Tres Montes dolmen is a megalithic monument, destroyed by a big fire and whose main frame of structure remains

**Figure 3.** General view of the room and corridor of Tres Montes (picture by J. Sesma Sesma).

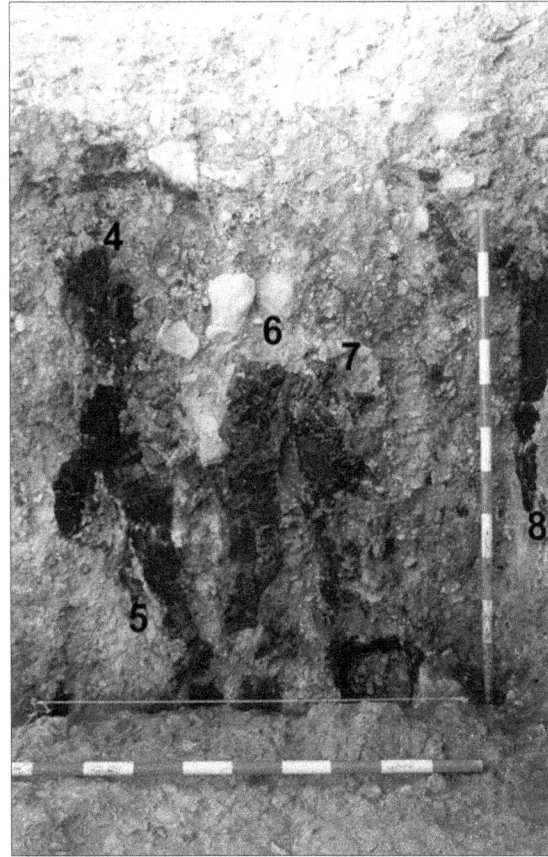

**Figure 4.** Detail of the perimetral wood posts in the room (picture by J. Sesma Sesma).

very well preserved; the preservation of carbonised wood offers a very valuable information about its construction, use and abandonment, and a direct evidence of constructive techniques in wood and stone.

The discovery of the dolmen occurred during the survey carried out in 1990 in the Bardenas Reales; the lack of dense vegetation and of human crops allowed the location of more than 400 archaeological sites (SESMA SESMA AND GARCÍA GARCÍA 1994); among them, Tres Montes offered very good perspectives of preservation. Therefore, in 1991 began the excavation of the monument (ANDRÉS RUPÉREZ ET AL. 2002: 198). The archaeological material, some atypical stone features and the paleonthological remains pointed to the existence of a burial layer (ANDRÉS RUPÉREZ ET AL. 1997, 2001).

The monument was composed of a room (of 4.20 x 3.40 m, and 2 m dug in the soil) and a corridor. A woody sustaining structure was documented along the walls, with 65 posts in a perimetral ditch, and other ones in the middle of the room, possibly sustaining the cover structure (FIG. 3) (ANDRÉS RUPÉREZ ET AL. 2001). Radiocarbon dates were obtained directly from timber: 4330 ± 110 BP and 4080 ± 100 BP (SESMA SESMA 1993).

Fire was the main preserving agent, since without the fire the whole woody structure would have decayed. Authors discuss about the intentionality of the fire, bacause this practice had already been documented in other sites as a part of burial ritual (ANDRÉS RUPÉREZ ET AL. 2001, 2002, MASSET 2002: 10, ROJO GUERRA ET AL. 2002: 21). In this paper, we present the results of the dendrological analysis of some of the perimetral posts of Tres Montes monument, those better preserved (some of them were completely burnt and only an ash mark remained on the wall) (FIG. 4).

Our analysis takes part of the general aims of Dendrology: the ecological, climatic and historical information from tree growth-rings. In archaeological charcoal, it is especially important to know about human techniques of construction, exploitation of vegetable landscape, and so on. Dendrological studies from prehistoric sites have recently multiplied, both on wood and charcoal, and the results are often succesful in spite of the frequent difficulties of studying this kind of materials (PÉTREQUIN 1989, 1997, PÉTREQUIN AND PÉTREQUIN 1989, BERNARD 1998, MARGUERIE 1992a, 1995a, b, MARGUERIE AND MARCOUX 2001, SAN JUAN ET AL. 1998, BOSCH ET AL. 2000).

After the botanical identification of the wood species, we could recognize some of their growing features under a low magnification observation of the samples. One of the bases of Dendrology is that normal growing rhythm of a plant can be altered by both internal and external agents, such as climatic conditions, human exploitation, animal grazing, microorganisms attack, etc. (MUNAUT 1988, MARGUERIE 1992a, SCHWEINGRUBER 1996). Some of the dendrological analysis criteria are the following ones:

- *Tree-ring width*, which is generally a direct reflect of the plant growth.

- *Tree-ring bending* can show the part of the tree where the sample comes from (HUNOT 2000: 12).

- *The presence of the pith and/or the bark*: this way, we can know the complete radius of the stem, and the felling season, depending on the bark position.

- *The presence of tyloses* in some deciduous species can mark the difference between the heartwood and the sapwood (ESAU 1985: 275).

- *Compression wood* is a reaction to non vertical elements, such as branches, bended stems and other changes in the stand of the tree (KAENNEL AND SCHWEINGRUBER 1995).

- *Radial cracks* seem to be a result of green wood burning, because there is a fast moisture loss, which makes wood contract (MARGUERIE 1992a-b, THÉRY-PARISOT 2001).

- *Vitrification of cellular tissues*: causes are not clear, but it could be related to the use of green wood and to a low-oxygen combustion (THÉRY-PARISOT 1998, 2001, FABRE 1996, TARDY 1998, THINON 1992, SCHEEL-YBERT 1998, CARRIÓN 2003).

- *The presence of fungi and xylophagous insects*: the most interesting discussion is about the state of the collected wood (dead, green, altered...) (THÉRY-PARISOT 1998, 2001).

- *The human work of wood* can be documented from polished, sharped, squared... surfaces. Only in some special archaeological contexts, wood tools are preserved (PETRÉQUIN 1989, 1997, PETRÉQUIN AND PETRÉQUIN 1989, BOSCH ET AL. 2000).

The tree ring measurement has been carried out throught a binocular lupe, with low magnification (between 8 and 32) on a measuring bench which allows to displace the sample with a precision of 0,01 mm. We have used the software TSAP (Time Series Analysis and Presentation) to build the database, not only of tree-rings width, but also of all the previously mentioned parameters and their graphic and statistic treatment.

All the samples had a similar size (between 10-15 cm diameter) and they had very narrow rings, so we have obtained quite long series. Several rays of each sample have been measured in order to get a curve as long as possible. The identification of all the samples as *Juniperus* hindered from the crossdating of the curves and from getting average values, because missing rings were very frequent. Unfortunately, we do not have any dendrochronological reference in our region of study in order to date the new obtained series and to correct the curves displacement.

## DENDROLOGICAL ANALYSIS RESULTS

### Botanical identification and microscopic analysis

The botanical identification of plant species is one of the bases of charcoal analysis. The observation is possible through a metallurgical microscope, between 100 and 500 magnifications (VERNET 1973, VERNET ET AL. 1979). The anatomical patterns of each species is compared with a reference collection of current carbonized woods and specialized bibliography (METCALFE 1960, METCALFE AND CHALK 1950, GREGUSS 1955, 1959, JACQUIOT 1955, JACQUIOT ET AL. 1973, SCHWEINGRUBER 1978, 1990, etc.). When a higher magnification is required, charcoal can be observed throught a Scanning Electron Microscope.

A single taxon was used in the timber of Tres Montes: *Juniperus* sp. (FIG. 5, 6 AND 7). Only the genus has been identified, but not the species, because of the enormous similarity of all of them. Greguss (1955) established a criterion for juniper species identification, according to the rays height. On the one hand, there is a group of species with short rays, up to 6 cells high (average of 2-4 cells), such as *Juniperus communis, J. excelsa, J. sabina* and *J. thurifera.* On the other hand, the species with higher rays (up to 12-15 cells) are *J. oxycedrus, J. phoenicea* and *J. phoetidissima.* In order to pun this criterion into a representative effect, the measurement of a high number of rays is necessary because of the intraspecific variability. Tres Montes charcoal was large enough to get very representative average values of the rays height. Those were 1 to 5 cells high (FIG. 6) so the samples belong to the short rays species. According to the current ecology in the region, the collected species might be *Juniperus thurifera.* The presence of this species points to an open vegetable landscape of supramediterranean stage, althought the selection of a single taxones for timber does not allow us to the further reconstruction of the plant formations. Moreover, it is impossible to know if this species was collected because of its abundance in the surrounding environment, or because of its wood qualities. Actually, juniper species have very twisted, gnarled trunks, but they can also produce straight ones under favourable growing conditions (NTINOU 2002).

**Figure 5.** Cross section of *Juniperus* sp. from Tres Montes.

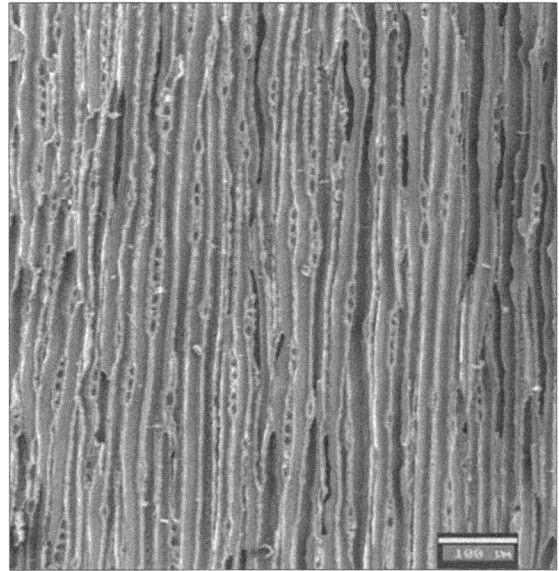

**Figure 6.** Tangential section of *Juniperus* sp. from Tres Montes.

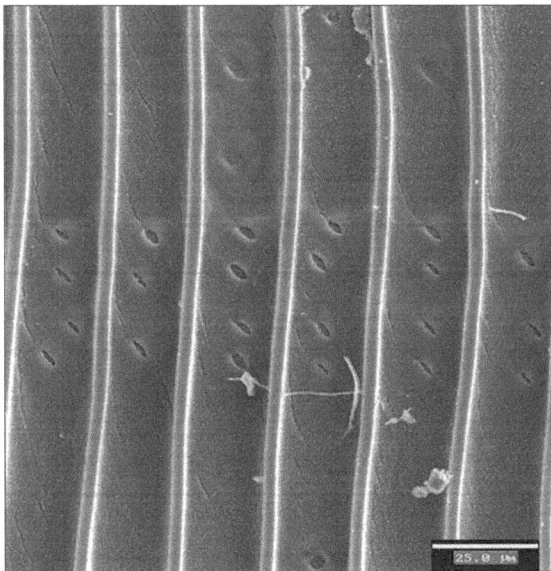

**Figure 7.** Radial section of *Juniperus* sp. from Tres Montes.

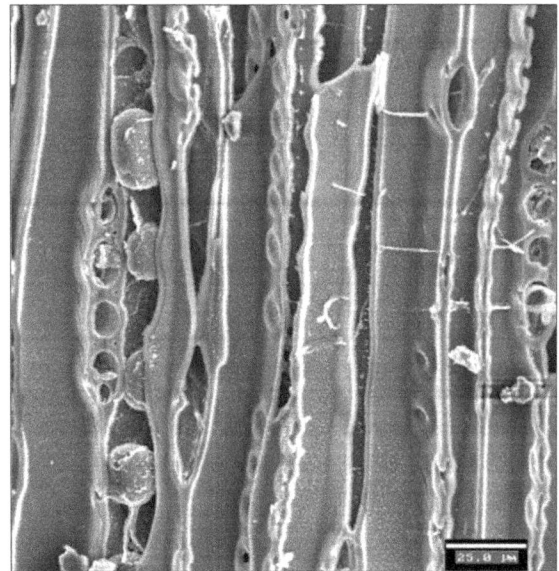

**Figure 8.** Detail of fungi hyphae on a *Juniperus* tangential

Microscopic analysis of Tres Montes charcoal has shown the presence of fungi hyphae in all the samples (FIG. 8). They are quite frequent in archaeological charcoal and wood (THÉRY-PARISOT 2001, CARRIÓN 2003, CARRIÓN AND BADAL 2004). Wood decay occurs under environmental factors favourable for microbial development (such as high moisture and oxygen levels). Fungi grow in the wood and feed on it, developping throught intervascular pits and along the vessel, until the whole structure is contaminated. Then, there is a gradual process of cohesion loss of the wood and cell walls become brittle and distructured. The most interesting discussion about fungi presence on charcoal treats of the state of the collected wood (THÉRY-

PARISOT 1998, 2001). Ethnographical studies have proved that dead wood was often collected for firewood (although contamination could also occasionally occur in living trees). Nevertheless, collected wood for timber is preferred to be as healthy as possible, especially if it is taking part of a sustaining structure.

In Tres Montes charcoal, fungi were present in 100% of the samples and charcoal fragments, and the contamination degree was very similar in all of them: the hyphae development was still incipient (FIG. 8). We think that contamination might have occurred after collection, while timber was taking part of the monument structure, because it is not very common to find the same contamination degree in several living trees. This way, timber long-term exposure to soil and air moisture may have helped the microorganisms development.

## Macroscopic analysis and growth-rings measurement

The timber of Tres Montes was very fragmented, but in a few cases it was possible to reconstruct the whole diameter of the trunk. That showed a selection of the wood size, between 10-15 cm. They did not appear to be hand-worked and the bark was never present. Several measures of 11 samples have been realized, those best preserved. Tree ring measurement was often difficult because *Juniperus* has not always visible growth-ring boundaries when looked under the binocular lupe. The presence of false rings makes also difficult the curves crossdating. This phenomenon is quite common in juniper species, but it has been also documented in other conifers, such as *Pinus halepensis, Pinus maritima, Pinus pinea, Pinus sylvestris, Pinus leucodermis, Pinus nigra* and *Larix decidua* (GUIBAL 1996: 507).

A frequent alteration in juniper anatomy is the formation of eccentric rings (ESPER 2000: 255) and it is generally related with the formation of compresion wood. This can be due to the growing of new shoots, to alterations on the growing position, to mechanical events, etc. (SCHWEINGRUBER 1996). Then, the measurement of several rays in the cross-section of the tree would offer very different values of the tree ring width. In Tres Montes, there is often a slightly excentricity of the pith, that we have tried to mitigate by measuring several rays of each sample. When reference dendrochronological series do exist, it is possible to correct the displacement of the curves due to the pressence of false and/or eccentric rings (WILES ET AL. 1999, ESPER 2000, BHATTACHARYYA ET AL. 1988, BILHAM ET AL. 1983, etc.). Since in our case there was no reference, the aims of the study were focused on ecological and human exploitation events, noticed from tree growth-rings.

First years growth-rings were present in most of the samples, but the pith was only preserved in sample TM12. The macroscopic observation of the samples showed a general low rate of growth, but this was quite irregular, with occasional large rings. Some rays have been measured in each sample, but we have not realized any average curve, because there was a slightly displacement on the curves belonging to a same tree, due to the presence of false rings.

The measurement results (FIG. 9) show the values of the tree rings width; some of them are quite short because we have only measured those parts of the sample where the rings boundaries were more visible. Then, we have

crossdated the curves from the same sample in order to get a series as long as possible. The samples generally have a very low rate of growth, with average values between 0.24 and 0.98 mm. We can observe a bimodal pattern of the values distribution: most of them are between 0.25 and 0.625, and a small gruop are between 0.75 and 1 mm.

When trying to crossdate the curves obtained from different samples, we notice the displacement of the main events because of the presence of false rings. According to the similarities in the main events of the curves, two dendrological series have been obtained. The series 1 (FIG. 10) is composed by curves from samples 6, 8, 18, 19 and 25. They present the highest growing values, and a very irregular tendency, with continuous rises and falls on the rate of growth. The discontinue lines mark

| SAMPLE | TAXON | RADIUS (mm) | No. RINGS | AVERAGE WIDTH |
|---|---|---|---|---|
| TM5-1 | *Juniperus* sp. | 34,17 | 92 | 0,37 |
| TM5-2 | *Juniperus* sp. | 20,05 | 50 | 0,40 |
| TM5-3 | *Juniperus* sp. | 30,55 | 70 | 0,44 |
| TM5-4 | *Juniperus* sp. | 22,53 | 74 | 0,30 |
| TM6-1 | *Juniperus* sp. | 25,76 | 52 | 0,50 |
| TM6-2 | *Juniperus* sp. | 24,55 | 62 | 0,40 |
| TM6-3 | *Juniperus* sp. | 13,77 | 24 | 0,57 |
| TM6-4 | *Juniperus* sp. | 24,95 | 61 | 0,41 |
| TM6-5 | *Juniperus* sp. | 7,14 | 15 | 0,48 |
| TM6-6 | *Juniperus* sp. | 19,64 | 44 | 0,45 |
| TM6-7 | *Juniperus* sp. | 4,1 | 12 | 0,34 |
| TM6-8 | *Juniperus* sp. | 17,56 | 38 | 0,46 |
| TM6-9 | *Juniperus* sp. | 27,67 | 64 | 0,43 |
| TM6-10 | *Juniperus* sp. | 15,4 | 29 | 0,53 |
| TM6-11 | *Juniperus* sp. | 25,52 | 47 | 0,54 |
| TM6-12 | *Juniperus* sp. | 14,75 | 17 | 0,87 |
| TM6-13 | *Juniperus* sp. | 33,91 | 70 | 0,48 |
| TM7-1 | *Juniperus* sp. | 43,05 | 155 | 0,28 |
| TM7-2 | *Juniperus* sp. | 12,24 | 39 | 0,31 |
| TM7-3 | *Juniperus* sp. | 30,47 | 120 | 0,25 |
| TM7-4 | *Juniperus* sp. | 20,77 | 54 | 0,38 |
| TM7-5 | *Juniperus* sp. | 22,09 | 50 | 0,44 |
| TM7-6 | *Juniperus* sp. | 24,19 | 100 | 0,24 |
| TM7-7 | *Juniperus* sp. | 29,66 | 99 | 0,30 |
| TM8-1 | *Juniperus* sp. | 22,75 | 67 | 0,34 |
| TM8-2 | *Juniperus* sp. | 40,05 | 107 | 0,37 |
| TM8-3 | *Juniperus* sp. | 22,08 | 61 | 0,36 |
| TM8-4 | *Juniperus* sp. | 50,26 | 125 | 0,40 |
| TM8-5 | *Juniperus* sp. | 15,87 | 50 | 0,32 |
| TM11-1 | *Juniperus* sp. | 23,07 | 79 | 0,29 |
| TM11-2 | *Juniperus* sp. | 29,59 | 107 | 0,28 |
| TM11-3 | *Juniperus* sp. | 29,57 | 93 | 0,32 |
| TM12-1 | *Juniperus* sp. | 23,09 | 45 | 0,51 |
| TM12-2 | *Juniperus* sp. | 18,86 | 37 | 0,51 |
| TM12-3 | *Juniperus* sp. | 3,81 | 12 | 0,32 |
| TM12-4 | *Juniperus* sp. | 21,99 | 52 | 0,42 |
| TM12-5 | *Juniperus* sp. | 15,59 | 29 | 0,54 |
| TM12-6 | *Juniperus* sp. | 18,96 | 44 | 0,43 |
| TM12-7 | *Juniperus* sp. | 22,33 | 52 | 0,43 |
| TM12-8 | *Juniperus* sp. | 8,31 | 19 | 0,44 |
| TM13-1 | *Juniperus* sp. | 15,39 | 52 | 0,30 |
| TM13-2 | *Juniperus* sp. | 13,49 | 55 | 0,25 |
| TM13-3 | *Juniperus* sp. | 18,45 | 65 | 0,28 |
| TM13-4 | *Juniperus* sp. | 16,04 | 50 | 0,32 |
| TM13-5 | *Juniperus* sp. | 11,89 | 42 | 0,28 |
| TM13-6 | *Juniperus* sp. | 14,75 | 54 | 0,27 |
| TM14-1 | *Juniperus* sp. | 24,24 | 28 | 0,87 |
| TM14-2 | *Juniperus* sp. | 27,21 | 29 | 0,94 |
| TM18-1 | *Juniperus* sp. | 21,46 | 26 | 0,83 |
| TM18-2 | *Juniperus* sp. | 28,78 | 42 | 0,69 |
| TM18-3 | *Juniperus* sp. | 40,31 | 46 | 0,88 |
| TM18-4 | *Juniperus* sp. | 21,91 | 24 | 0,91 |
| TM18-5 | *Juniperus* sp. | 17,63 | 20 | 0,88 |
| TM19-1 | *Juniperus* sp. | 21 | 22 | 0,95 |
| TM19-2 | *Juniperus* sp. | 19,5 | 20 | 0,98 |
| TM19-3 | *Juniperus* sp. | 22,14 | 27 | 0,82 |
| TM19-4 | *Juniperus* sp. | 19,14 | 23 | 0,83 |
| TM19-5 | *Juniperus* sp. | 16,78 | 20 | 0,84 |
| TM25-1 | *Juniperus* sp. | 30,01 | 49 | 0,61 |
| TM25-2 | *Juniperus* sp. | 26,19 | 44 | 0,60 |
| TM25-3 | *Juniperus* sp. | 28,09 | 48 | 0,59 |
| TM25-4 | *Juniperus* sp. | 25,7 | 56 | 0,46 |
| TM25-5 | *Juniperus* sp. | 35,87 | 79 | 0,45 |

**Figure 9.** Results of the growth-ring measurement of tres Montes samples.

some growing events noticed in all the curves, which have helped to the general crossdating, but make clear a displacement in some of them.

Series 2 (FIG. 11) is composed by samples 5, 7, 12, 13 and 14. Curves of sample TM12 are very irregular, probably due to the pith presence, because trees growing is very

irregular and more sensitive to external events during the first years of the plant. The longest curves show very low growing rhythm, especially obvious in TM7-1, TM7-3, TM7-6, TM8-2, TM8-4 and TM11-2, which exceed 100 rings/years. Tendency of TM13, with a sequence of continuous rises and falls of the growth has frequently been thought to be a characteristic growth response of plants to the human exploitation (MARGUERIE 1992a).

On the other hand, the gradual decrease of the growth, obvious in most of the individuals (especially in TM7 and TM5) can be due to climatic causes. According to other authors, the precipitation levels are one of the main

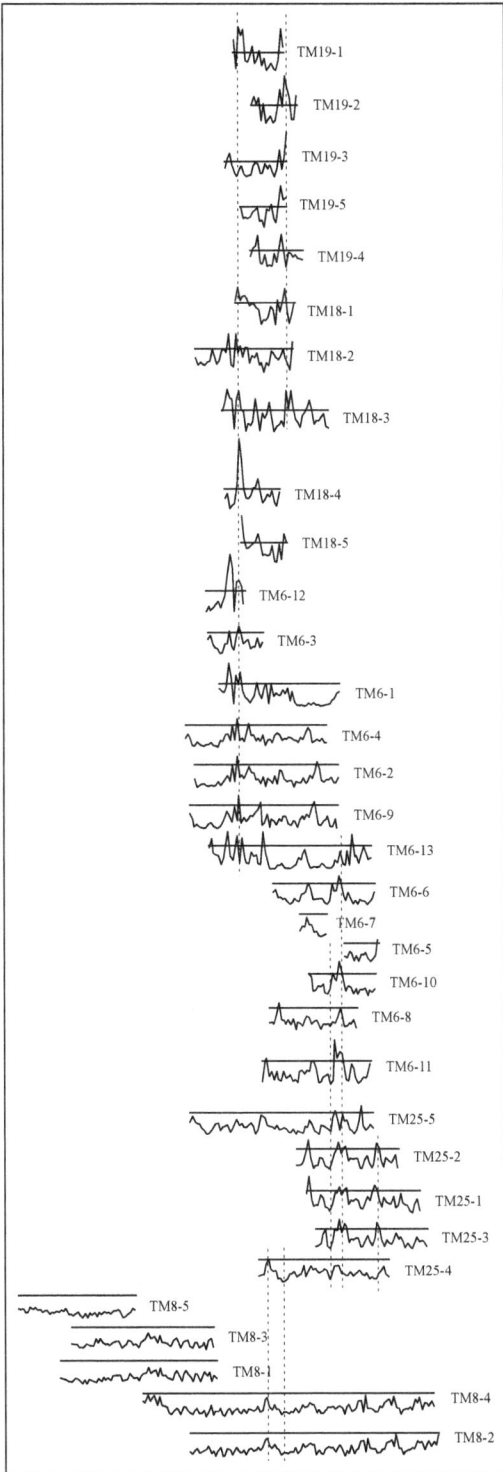

Figure 10. Dendrological series 1.

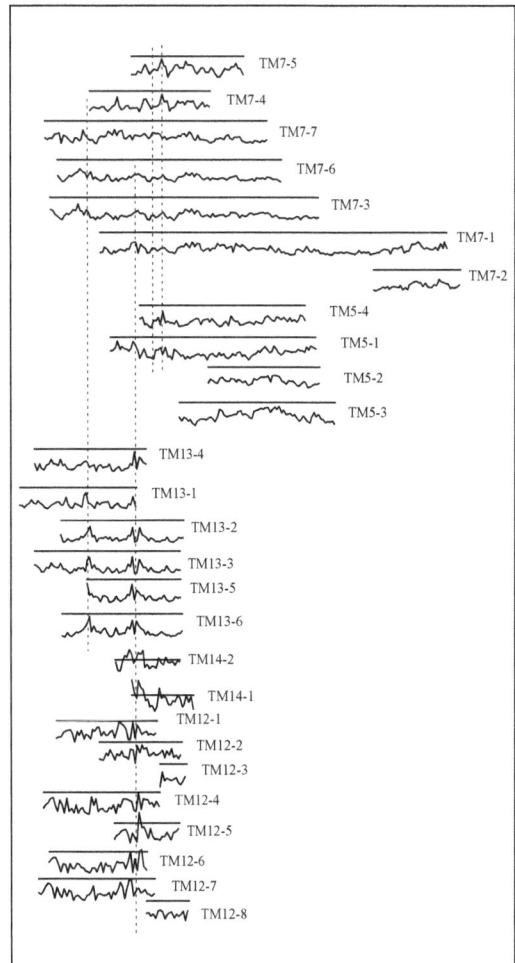

Figure 11. Dendrological series 2.

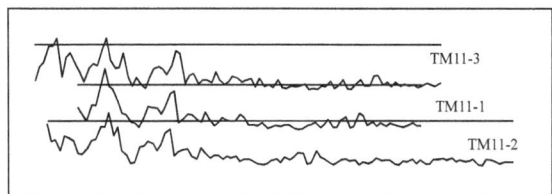

Figure 12. Dendrological curve of sample TM11.

factors causing a very low rate of growth in conifers. A dendrological study carried out by Esper (2000) on a *Juniperus* population in Pakistan offered growing values between 0.24 and 0.42 for high mountain individuals, in contrast to those living at the bottom of the valley, with a higher rate of growth. The sensibility to the lack of rain gets more obvious in the long curves. Moreover, junipers are also very sensitive to extreme temperatures (ESPER 2000: 259). The Bardenas Reales are actually characterized by the heterogeneous precipitation (long drought periods and occasional, torrential rains) and great temperature contrasts (up to 45°C oscillations from day to night) (ELÓSEGUI AND URSÚA 1994: 17-19), so these climatic patterns may have influenced plants growth.

Finally, curves from sample TM11 could not be correlated with anyone else (FIG. 12). They present a high variability in the rate of growth near the pith, and a gradual decrease to the end of the curves.

## *JUNIPERUS* MANAGING AND ITS ROLE IN TRES MONTES ENVIRONMENT

The collection of a single taxon, *Juniperus* sp. for Tres Montes timber makes clear a selection of this species maybe because of both, the physical and mechanical wood qualities or its abundance in the surrounding area. Generally speaking, timber is selected according to the use it is meant for: sometimes straight trunks are needed; other ones, flexible branches or stems are plaited for roofs and walls. Endurance and resistance to moisture, to open air or to fungal attack are very appreciated wood qualities for timber. In Tres Montes, there is also a selection in the size of the trunks: those of the same diameter could have been obtained from trees of similar age. Juniper species are not exactly tree-like, because they have very often several branches from the base. Among this genus, *Juniperus thurifera* generally presents a single, straight trunk from the base. Nevertheless, some studies carried out on current *Juniperus thurifera* rates of growth show a bushy structure, with some trunks, all of them growing in a synchronic way (with a similar number of rings) (BERTAUDIÈRE *ET AL*. 2001). Those plants would be very useful for getting timber with a similar diameter, because some studied individuals produced more than 10 branches from the base. In that case, the general biomass is higher than in those with a single trunk. Current human groups in Morocco exploit *Juniperus thurifera* populations for several tasks, such as to obtain timber, and this practice has made evident the great regeneration capacity of this species after the continuous human managing (BARBERO *ET AL*. 1990).

In tres Montes, the similarity among some curves from different samples could be due to the use of several branches from a same individual, although it is impossible

to know from the available data. Is has been proved that juniper ramification can be a reaction to a severe environment and to an intense human exploitation. In our case, both hypotheses had already been posed from dendrological results.

The absence of other taxa in Tres Montes charcoal do not let us approach to the plant formations near the site. The use of juniper makes obvious its presence in the area, but it would be very interesting to know its percentage or the presence-absence of other arboreal species, because if shrub-like formations were in dominance, junipers might be collected just because of the lack of other trees for timber.

Palaeobotanical studies can offer a helpful approach to these questions but, unfortunately, those are rare in this region. Pollen analysis carried out in the Bronze Age sites of Puy Aguila I and Monte Aguilar, in the Bardenas Reales, show a landscape dominated by herbaceous and shrub species, characteristic of arid environments (Chenopodiaceae, Compositae, Poaceae, etc.) and very few trees (IRIARTE 1992, 2001). The layers previous this habitation show more abundant tree vegetation, with *Quercus* t. *ilex-coccifera* and *Pinus*, and the highest percentage of Cupressaceae (which the genus *Juniperus* belongs to). Effects of tree felling become clearer in pollen sequences during Recent Prehistory, because pines and river species seemed to be more abundant before Bronze Age occupation (IRIARTE 1992). Reduction of arboreal vegetation occurs in parallel with to Cerealia curves progression. This fact makes obvious the human causes of the forest disappearance. According to this idea, an interesting point is the gradual regeneration of pines where cereal crops are abandoned, so this genus must have a basic role in the potential regional vegetation. Then, why did not Bell-Baker groups exploit pines for Tres Montes construction? Maybe, the species selection was not focused on the environmental availability, but on the physical and mechanical qualities of the wood, according to its future use.

## CONCLUSION

Dendrological analysis of the charcoal from Tres Montes has made evident some methodological difficulties in conifer growth-rings measurement. The presence of false rings becomes clear in the curves displacement. This does not allow us to perfectly crossdating the series, even from a same individual. The existence of regional series or a high number of samples from a same site would help to correct the curves, but dendrological analysis in Spain are still incipient, so we have no reference for this region. In spite of these limits, we can offer some interesting conclusions.

All the curves obtained from the same sample present a

quite good crossdating (despite displacement due to false rings). But the absolute values of the growth-rings width from different samples are variable, so they seem to come from several individuals. Those differences might have been highlighted by eccentricity in the tree cross section.

Two dendrological series have been obtained (only sample TM11 has no correlation) which could belong to a two different plant formations, exploited for timber. The rate of growth is generally very low. This fact can be due to a strong climatic sensibility. In fact, this is a common growth reaction of junipers to a low precipitation regime and to temperature contrasts. It has been also documented in other regions, where growth values are similar to those of Tres Montes. Then a severe climate (low precipitations, long drought periods, large temperature oscillations) could

be inferred in the Bardenas Reales for Bell-Baker period, as it is today. Moreover, there might be an intense human exploitation of forests, because some dendrological curves show a characteristic pattern of this activity: a succession of dramatic falls in the growth (corresponding to the years of exploitation) and quick rises (following the exploitations there are some years of growing recovering).

In conclusion, the selection of juniper wood for timber procurement in Tres Montes seems to be focused on this species because of both, its inner qualities for construction and/or its availability in the environment near of the site.

REFERENCES

ANDRÉS RUPÉREZ M.T., GARCÍA GARCÍA M.L., SESMA SESMA J., 1997.- El sepulcro calcolítico de Tres Montes (Las Bardenas Reales, Navarra), in: R. De Balbín Behrmann and P. Bueno Ramírez (eds.), Neolítico, Calcolítico y Bronce (Tomo II). Actas del II Congreso de Arqueología Peninsular (Zamora, del 24 al 27 de Septiembre de 1996), Madrid, Alcala de Henares: 301-308.

ANDRÉS RUPÉREZ M.T., GARCÍA GARCÍA M.L., SESMA SESMA J., 2001.- El sepulcro campaniforme de Tres Montes (Las Bardenas Reales, Navarra). Intervención de urgencia de 1991 y campañas de 1996 y 1997, Trabajos de Arqueología de la Universidad de Navarra, 15: 315-322.

ANDRÉS RUPÉREZ M.T., GARCÍA GARCÍA M.L., SESMA SESMA J., 2002.- Una tumba destruida por el fuego: el sepulcro campaniforme de Tres Montes, en las Bardenas Reales (Navarra), in: M. Rojo Guerra and M. Kunst (eds.), El significado del fuego en los rituales funerarios del Neolítico, Valladolid, Studia Archaeologia, 91: 191-218.

BARBERO M., QUEZEL P., LOISEL R., 1990.- Les apports de la phytoécologie dans l'interprétation des changements et perturbations induits par l'homme sur les écosystèmes forestiers méditerranéens, Forêts méditerranéennes, XII: 193-215.

BERNARD V., 1998.- L'homme, le bois et la forêt dans la France du nord entre le Mésolihique et le haut Moyen Age. Oxford, BAR Publishing, 190 p. (BAR International Series 733).

BERTAUDIÈRE V., MONTÈS N., BADRI W., GAUQUELIN T., 2001.- La structure multicaule du genévrier thurifère: avantage adaptatif à un environnement sévère?, Comptes Rendus de l'Académie des Sciences, Series III-Sciences de la Vie, 324 (7): 627-634.

BHATTACHARYYA A., LA MARCHE V.C., TELEWSKI F.W., 1988.- Dendrochronological reconaissance of the conifers of northwestern India, Tree-Ring Bulletin, 48: 21-30.

BILHAM R., PANT G.B., JACOBY G.C., 1983.- Dendroclimatic potential of Juniper trees from the Sir Sar range in the Karakoram, Man and Environment, 7: 45-50.

BOSCH I LLORET A., CHINCHILLA SÁNCHEZ J., TARRÚS I GALTER J. (eds.), 2000.- El poblat lacustre neolític de La Draga. Excavacions de 1990 a 1998, Girona, Museu d'Arqueologia de Catalunya, Centre d'Arqueologia Subaquàtica de Catalunya, 296 p. (Monografies del CASC 2).

CARRIÓN Y., 2003.- Afinidades y diferencias de las secuencias antracológicas en las vertientes mediterránea y atlántica de la península Ibérica, Tesis Doctoral, Universitat de València, 572 p.

CARRIÓN Y., BADAL E., 2004.- La presencia de hongos e insectos xilófagos en el carbón arqueológico. Propuestas de interpretación, in: M.J. Feliu Ortega, J. Martín Calleja, M.C. Edreira Sánchez, M.C. Fernández Lorenzo, M.P. Martínez Brell, A. Gil Montero and R. Alcántara Puerto (eds.), Avances en Arqueometría 2003, Cádiz, Servicio de Publicaciones de la Universidad de Cádiz: 98-106.

ELÓSEGUI ALDASORO J., URSÚA SESMA C., 1994.- Las Bardenas Reales, Pamplona, Fondo de publicaciones del Gobierno de Navarra, 63 p.

ELÓSEGUI ALDASORO J., URSÚA SESMA C., ARBILLA IBÁÑEZ J., 1990.- Las Bardenas Reales. Mapa escala 1:50.000. Servicio Geográfico del Ejército. Gobierno de Navarra.

ESAU K., 1985.- Anatomía vegetal, Barcelona, Editions Omega, 779 p.

ESPER J., 2000.- Long-term tree-ring variations in Juniperus at the upper timber-line in the Karakorum (Pakistan), The Holocene, 10 (2): 253-260.

FABRE L., 1996.- Le charbonnage historique de la chênaie à Quercus ilex L. (Languedoc, France): conséquences écologiques, Doctorat, Université Montpellier II, 446 p.

GREGUSS P., 1955.- Identification of Living Gymnosperms on the Basis of Xylotomy, Budapest, Akadémiai Kiado, 263 p.

GREGUSS P., 1959.- Holzanatomie der Europaïschen Laubhölzer und Sträucher, Budapest, Akadémiai Kiado, 330 p.

GUIBAL F., 1996.- Dendrochonological studies in the french mediterranean area, in: J. Dean, D.M. Meko and T.W.

Swetnam (eds.), *Tree rings, environment and humanity. Proceedings of the International Conference (Tucson, Arizona, 17-21 May 1994)*, Tucson, Department of Geosciences, University of Arizona: 505-513.

HUNOT J.-Y., 2000.- *Les restes de bois carbonisés provenant de constructions médiévales angevines*, Mémoire de DEA, Université de Rennes 2, 40 p.

IRIARTE M.J., 1992.- El entorno vegetal en las Bardenas Reales (Navarra) durante la Prehistoria reciente, *Cuadernos de Sección, Historia*, 20: 359-367.

IRIARTE M.J., 2001.- Un caso paradigmático de antropización del medio vegetal. El poblado de la Edad del Bronce de Puy Águila I (Bardenas Reales, Navarra), *Trabajos de Arqueología Navarra*, 15: 123-136.

JACQUIOT C., 1955.- *Atlas d'anatomie des bois des conifères*, Paris, Centre Technique du Bois, 133 p.

JACQUIOT C., TRENARD Y., DIROL D., 1973.- *Atlas d'anatomie des bois des angiospermes (Essences feuillues)*, Paris, Centre Technique du Bois, 175 p.

KAENNEL M., SCHWEINGRUBER F. H., 1995.- *Multilingual Glossary of Dendrochronology*, Bern, Verlag Paul Haupt, 467 p.

MARGUERIE D., 1992a.- *Evolution de la végétation sous l'impact anthropique en Armorique du Mésolithique aux périodes historiques*, Rennes, Editions U.P.R. n° 403 du C.N.R.S., 313 p. (Travaux du Laboratoire d'Anthropologie de Rennes, 40).

MARGUERIE D., 1992b.- *Charbons de bois et paléoenvironnement atlantique*, Rennes, Laboratoire d'Anthropologie de l'Université de Rennes, 1: 15-19 (Les bois archéologiques, dossier n° 2 A.G.O.R.A.).

MARGUERIE D., 1995a.- L'état du milieu forestier durant la Protohistoire et l'Antiquité en Bretagne. L'apport de l'anthracologie, *in*: J.-C. Beal (ed.), *L'arbre et la forêt, le bois dans l'Antiquité*, Paris, Publications de la Bibliothèque Salomon-Reinach, VII: 27-33.

MARGUERIE D., 1995b.- Paléoenvironnement des monuments mégalithiques de Saint-Just. Les études archéobotaniques, *in*: J. Briard, M. Gautier and G. Leroux (eds), *Les mégalithes et les tumulus de Saint-Just (Ile-et-Vilaine): évolution et acculturation d'un ensemble funéraire*, Paris, Editions du C.T.H.S.: 128-142.

MARGUERIE D., MARCOUX N., 2001.- Environnement des sépultures de l'Age du Bronze de Balchoĭkazakhbaĭevo (Oural, Russie), *Revue d'Archéologie de l'Ouest*, supplément n° 9: 241-243.

MASSET C., 2002.- Ce qu'on sait, ou croit savoir, du rôle du feu dans les sépultures collectives néolithiques, *in*: M. Rojo Guerra and M. Kunst (eds.), *El significado del fuego en los rituales funerarios del Neolítico*, Valladolid, Studia Archaeologia, 91: 9-20.

METCALFE C.R., 1960.- *Anatomy of Monocotyledons, I. Gramineae*, Oxford, The Clarendon Press, 731 p.

METCALFE C.R., CHALK L., 1950.- *Anatomy of Dycotiledons*. Oxford, Clarendon Press, 2 vol.

MUNAUT A.-V., 1988.- Les cernes de croissance des arbres (la dendrocrhonologie), *in*: L. Genicot (ed.), *Typologie des sources du Moyen-Age occidentale*, Turnhout-Belgium, Brepols, B III-2 (53): 1-51.

NTINOU M., 2002.- *El paisaje en el norte de Grecia desde el Tardiglaciar al Atlántico. Formaciones vegetales, recursos y usos*, Oxford, BAR Publishing, 268 p. (BAR International Series 1038).

PÉTREQUIN P. (dir.), 1989.- *Les sites littoraux Néolithiques de Claivaux-les-Lacs (Jura). II Le Néolithique Moyen*, Paris, Editions de la Maison des Sciences de l'Homme, 508 p.

PÉTREQUIN, P (dir.), 1997.- *Les sites littoraux Néolithiques de Claivaux-les-Lacs et de Chalain (Jura). III Chalain Station 3. 3200-2900 av. J.-C.* Vol. 1-2, Paris, Editions de la Maison des Sciences de l'Homme, 765 p.

PÉTREQUIN P., PÉTREQUIN A-M., 1989.- *Habitat lacustre de Bénin. Une approche ethno-archéologique*, Paris, Editions Recherche sur les Civilisations, 214 p.

ROJO GUERRA M., KUNST M., PALOMINO LÁZARO A.L., 2002.- El fuego como procedimiento de clausura en tres tumbas monumentales de la submeseta norte, *in*: M. Rojo Guerra and M. Kunst (eds.), *El significado del fuego en los rituales funerarios del Neolítico*, Valladolid, Studia Archaeologia, 91: 21-38.

SAN JUAN G., DRON J.-L., ARBOGAST R.-M., BORTUZZO L. ET AL. 1998. Le site néolithique moyen de Derrière-les-Près à Ernes (Calvados), *Gallia Préhistoire*, 39: 151-237.

SCHEEL-YBERT R., 1998.- *Stabilité de l'écosystème sur le littoral Sud-Est du Brésil à l'Holocène Supérieur (5500-1400 ans BP). Les pêcheurs-cueilleurs-chasseurs et le milieu végétal: apports de l'anthracologie*, Doctorat, Université de Montpellier II, 510 p.

SCHWEINGRUBER F.H., 1978.- Anatomie microscopique du bois, Birmensdorf, Editions de l'Institut fédéral de recherches sur la forêt, la neige et le paysage, 226 p.

SCHWEINGRUBER F.H., 1990.- *Anatomie europäischer Hölzer*, Bern/Stuttgart, Verlag Paul Haupt, 800 p.

SCHWEINGRUBER F.H., 1996.- *Tree Rings and Environment. Dendroecology*, Bern/Stuttgart/wien, Verlag Paul Haupt, 609 p.

SESMA SESMA J., 1993.- Aproximación al problema del hábitat campaniforme, El caso de las Bardenas Reales de Navarra, *Cuadernos de Arqueología de la Universidad de Navarra*, 1: 53-119.

SESMA SESMA J., GARCÍA GARCÍA M.L., 1994.- La ocupación desde el Bronce Antiguo a la Edad Media en las Bardenas Reales de Navarra, *Cuadernos de Arqueología de la Universidad de Navarra*, 2: 89-218.

TARDY C., 1998.- Anthracologie, *in*: S. Vache, S. Jérémy and J. Briand (eds.), *Amérindiens du Sinnamary (Guyane). Archéologie en forêt équatoriale*, Paris, Editions de la Maison des Sciences de l'Homme: 94-102 (Documents d'Archéologie Française, 70).

THÉRY-PARISOT I., 1998.- *Economie du combustible et paléoécologie en contexte glaciaire et périglaciaire, Paléolitique Moyen et Supérieur du Sud de la France. Anthracologie, Expérimentation, Taphonomie*, Doctorat, Université de Paris I, 499 p.

THÉRY-PARISOT I., 2001.- *Economie des combustibles au*

*Paléolitique. Expérimentation, taphonomie, anthracologie.* Editions C.N.R.S. - C.E.P.A.M., 195 p. (Dossier de Documentation Archéologique, 20).

THINON M., 1992.- *L'analyse pédoanthracologique: aspects méthodologiques et applications,* Doctorat, Université d'Aix-Marseille III, 317 p.

VERNET J.-L., 1973.- *Etude sur l'histoire de la végétation du sud-est de la France au Quaternaire, d'après les charbons de bois principalement,* Montpellier, Université des Sciences et Techniques, Laboratoire de paléobotanique: 81-90 (Paléobiologie Continentale, 4 (1)).

VERNET J.-L., BAZILE E., EVIN J., 1979.- Coordination des analyses anthracologiques et des datations absolues sur charbons de bois, *Bulletin de la Société Préhistorique Française,* 76 (3): 76-79.

WILES G.C., POST A., MULLER E.H., MOLNIA B.F., 1999.- Dendrochronology and Late Holcene History of Bering Piedmont Glacier, Alaska, *Quaternary Reasearch,* 52: 185-195.

# WOOD-ANATOMICAL EVIDENCE OF POLLARDING IN RING POROUS SPECIES: A STUDY TO DEVELOP?

Stéphanie THIEBAULT

MAE, Protohistoire européenne, Archéobotanique et paléo-écologie
UMR 7041 ArScAn CNRS, Université de Nanterre Paris X
21 allée de l'Université F- 92023 Nanterre (France)
stephanie.thiebault@mae.u-paris10.fr

ABSTRACT: Tree fodder (leaves, twigs) to feed ruminants seems to have been used since the Neolithic, although there is little direct archaeological evidence. The recognition of "ecological anomalies" in charcoal diagrams of cave sheepfolds, identified in the South of France, constitutes an additional indication. They suggest that certain species such as ash and deciduous oak were selected and gathered to feed animals during seasons of deficiencies, or as food complement. We have now to prove, by wood-anatomical study of wood charcoal from cave sheepfolds, the existence of pollarding.
KEY WORDS: Neolithic, France, charcoal fragments, tree fodder, pollarding

RÉSUMÉ: L'utilisation du fourrage d'arbre, depuis le Néolithique, est prouvée sur les sites favorables comme les gisements lacustres suisses. Cependant, peu d'éléments directs étayent l'hypothèse selon laquelle ce mode de nourrissage a été utilisé à cette époque dans les grottes du Sud de la France notamment. La mise en évidence "d'anomalies écologiques" dans les diagrammes anthracologiques des grottes bergeries du Sud de la France constitue un indice supplémentaire. Elles suggèrent que certaines espèces, comme le frêne, les chênes à feuillage caduc par exemple, étaient choisies et récoltées pour le nourrissage des animaux pendant les saisons de carence ou comme complément alimentaire. Il nous faut maintenant prouver, par l'étude anatomique des fragments de charbons issus des niveaux de bergerie, l'existence de stigmates d'émondage si celui-ci était pratiqué.
MOTS-CLÉS: Néolithique, France, charbon de bois, fourrage d'arbre, stigmate d'émondage

ZUSAMMENFASSUNG: Laubfutter (Laub, Äste) wurde vermutlich seit dem Neolithikum für Wiederkäuer genutzt, obwohl wenige direkte archäologische Beweise vorliegen. In Südfrankreich wurden Holzkohlen aus einer als Schafunterstand benutzten Höhle untersucht. Die Holzkohlen-Diagramme weisen "ökologische Anomalien" auf, die einen Hinweis auf Laubfütterung liefern. Sie legen nahe, dass bestimmte Arten wie Eschen und sommergrüne Eichen gesammelt wurden, um die Tiere während Notzeiten oder als Ergänzung zu füttern. Durch holzanatomische Studien der Holzkohlen aus dieser Höhle soll gezeigt werden, dass die Bäume geschneitelt wurden (*Translation Petra Zibulki*).
STICHWORTE: Tierfutter, Laubfutter, Neolithikum, Frankreich, Holzkohlen, Schaf, Höhle, Schneiteln

The history of forest management is a theme that has been developed since some years in archaeology, most particularly with the expansion of anthracology (DURAND 1992, CHABAL 1997, VERNET 1997, DUFRAISSE 2002, LUDEMANN 2002, NELLE 2002, PIQUÉ AND BARCELO 2002, etc.).

The progress of the disciplines allows today to advance a hypothesis according to which the forest management dates back to at least the Neolithic. The study of the traditional societies supplied models on the different uses of the products of trees (DURAND-TULLOU 1972, HALSTEAD 1998)

often supported by the written sources, dating mostly from the Middle Ages (BECHMANN 1984, BREICHER ET AL. 2002).

It seems, indeed, that the forest was used for numerous purposes. Several authors (HAEGGSTROM 1992, AUSTAD 1988) described the economical interests of the various parts of the tree. In this article we shall focus particularly on the small branches obtained by pruning. If the pollarded trees are recognizable by a characteristic form (FIG. 1) the treatment that they undergo also leaves visible marks (HAAS AND SCHWEINGRUBER 1994) in their wood anatomy.

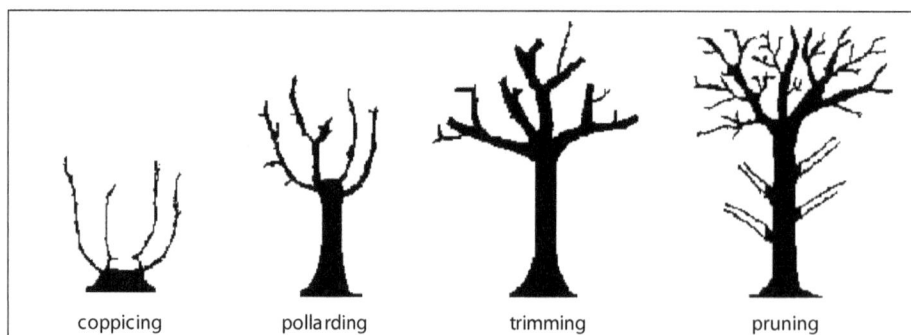

**Figure 1.** Characteristic forms of trees after different modes of cutting.

## WOOD POLLARDING AN ANCIENT TECHNIQUE

Usually, in the grazed woodland the small branches are harvested every three/four or six years. They have two main uses: firewood and animal fodder.

Since the beginning of breeding, animal food constituted a major preoccupation, especially during the seasons of scarcity (winters or dry summers). Until recent times, and even today in certain traditional societies, the breeders use leaves and twigs for fodder (DURAND-TULLOU 1972, AUSTAD 1988, HAAS ET AL. 1998, HALSTEAD 1998).

The means and the techniques used by the first European breeders are, for the moment, little known and the direct archaeological witnesses relating to it remain hardly accessible, except in the favorable deposits from lakeside settlements as Egolzwil, Weir or Arbon Bleiche in Switzerland (RASMUSSEN 1989, 1993, AKERET AND JACOMET 1997, AKERET ET AL. 1999; AKERET AND RENTZEL 2001). However, we advance the hypothesis according to which this mode of feeding has been used since the Neolithic in the major part of Europe and in particular, according to our own results, in the South of France.

## HOW TO HYPOTHESIZE THE PRESENCE OF CERTAIN SPECIES FOR ANIMAL FOOD IN ANTHRACOLOGICAL DIAGRAMS?

The results of palaeobotanical analyses (charcoal and pollen analyses) realised during the last years, on sediments from archaeological deposits of open air settlements and caves sheepfolds, bring elements of thought. They indicate that certain species (ash tree, deciduous oak, maple, hazel tree, elm...) were preferentially selected and harvested to feed animals, either during the seasons of deficiencies, or as a food complement.

The various plant associations and their dynamics are today well known thanks to the numerous data provided by charcoal analysis. They present a very good phytosociological coherence (leading species, association, vegetal succession...) during the Holocene. Sometimes the results of charcoal analysis unperfectly fit in to the general pattern. In this way we define in an anthracological diagram an "ecological anomaly" as an over-representation of a species (or of several species) which, usually, does nor belong, or very little, to the plant association in which the wood is harvested, neither to the corresponding ecological dynamics (replacements, reconquests). This suggests that there was a managed harvest of this (or of these) species in a precise goal. This over-representation is obvious either when the index of Pareto's concentration (usually of 20/80) shows a strong concentration (10/90 for example) or when the species seems abnormal among the association otherwise described (CHABAL 1997).

The first anomaly had been revealed in the anthracological diagram of Coufin 2 (Isère)(FIG. 2). The anthracological diagram shows two curves in *ante coincidence*: the curve of yew (*Taxus baccata*) and of ash tree (*Fraxinus excelsior*) (THIÉBAULT 1988). The curve of yew is divided in to four parts. The first, contemporary with the Mid Neolithic (F10 to F9), corresponds to the domination of yew, developed under a wet mountain climate. During the second part, from the Chalcolithic to the final Bronze Age 1 (from F8 to F4), the curve of yew is clearly decreasing to the benefit of deciduous oak (*Quercus* f.c.), maple (*Acer* sp.), ash tree, hazel tree (*Corylus avellana*) and beech (*Fagus sylvatica*) to a lesser extent. It could be due to a change in the climatic conditions resulting from the increase of the nebulousness. From the final Bronze Age (F4 to F3a) to the first Iron Age, the yew reaches again important percentages and finally disappears during late Antiquity (F2a and F1).

These results were compared to those of the analysis of faunal remains from the same levels. Their identification by B. Caillat (in BINTZ 1981) show that sheep and goat dominate in the layers where the yew is majority, that the cattle remains dominate from

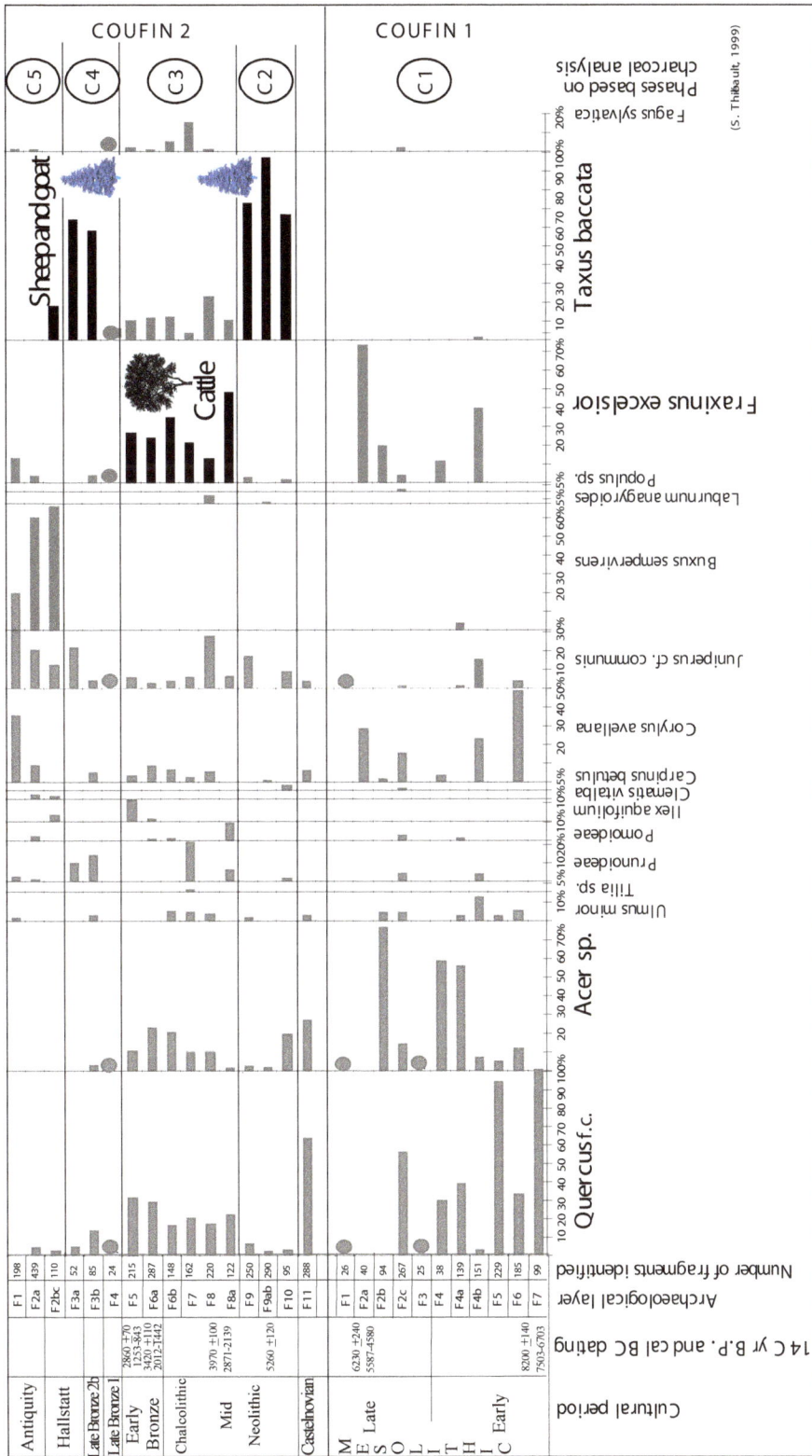

**Figure 2.** Anthracological diagram of Coufin 2 (Isère, France). On the abscissa appear the percentages of the number of fragments identified for each taxa, the ordinate presents the archaeological level and the number of fragments identified.

layer F8 to layer F6, in which the ash tree is the most important species. They are replaced by those of sheep and goat in the following layers where, again, the yew replaces the ash tree. Whatever is the cause of the temporary decrease of the cattle, the breeding of these animals is synchronous with an increase of the curve of the ash tree and the decrease of the yew, which is a very

poisonous species. One of the hypotheses proposed is that the over-representation of the ash tree would result from the harvest of this species for feeding the bovins penned in the rock shelter. On the contrary sheep and goat, grazing in natural meadows, would not be dependent on additional food brought to this site (THIÉBAULT 1988).

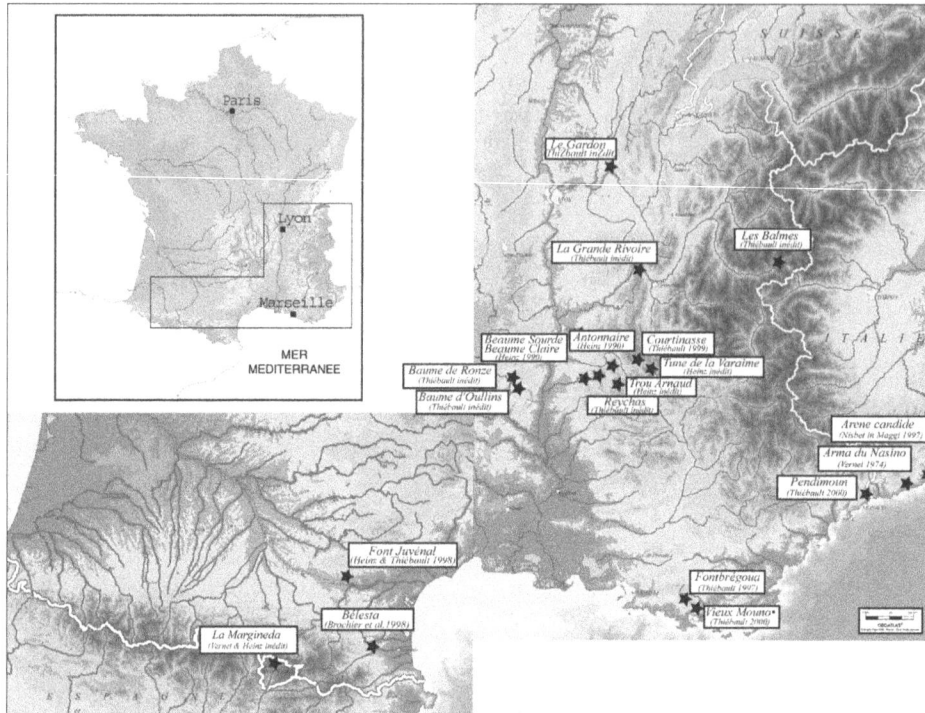

**Figure 3.** Map of the sheepfold caves in the South of France analysed by anthracology.

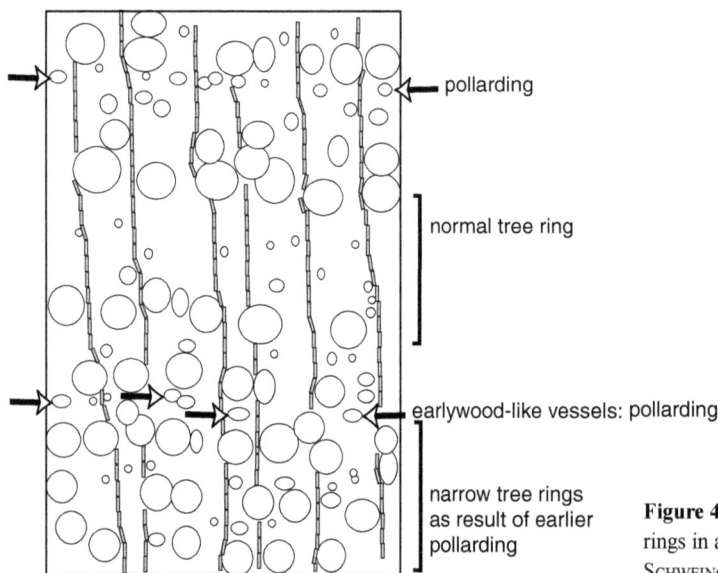

**Figure 4.** Schematised sequence of narrow and wide tree rings in a pollarded ash cross section (from HAAS AND SCHWEINGRUBER 1994: 37).

Since the study of charcoal from Coufin 2, about twenty sheepfold caves have been the object of an anthracological analysis (FIG. 3). "Ecological anomalies" were observed or suggested for at least eight published archaeological sites (THIÉBAULT IN PRESS). The sites in question are: Arene Candide in Italy (NISBET 1997, THIÉBAULT 2001); in France, Fontbrégoua and the Vieux Mounoï in the Var (THIÉBAULT 1997); Antonnaire and the Baume de Courtinasse in the Drôme, (HEINZ in ARGANT ET AL. 1991, THIÉBAULT 1999); Baume d'Oullins, in Ardèche (THIÉBAULT unpublished) and Bélesta, in the Pyrénées Orientales (HEINZ in BROCHIER ET AL. 1998).

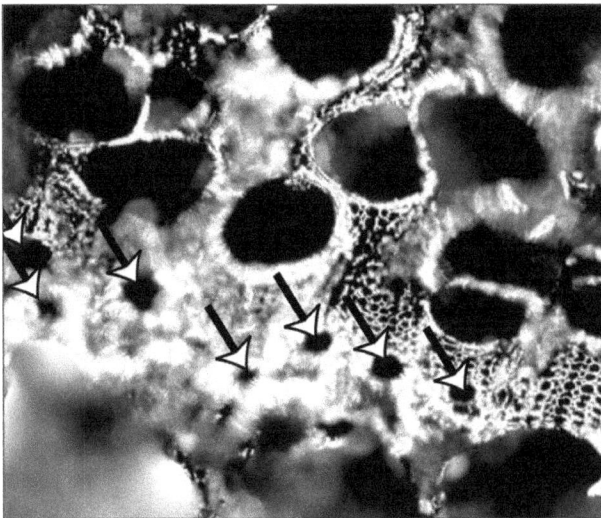

**Figure 5.** Earlywood-like vessels in the latewood of an ash tree (*Fraxinus excelsior*) from a Mid Neolithic layer of the Gardon cave (Ain, France), (cross section, enlargement: 100x).

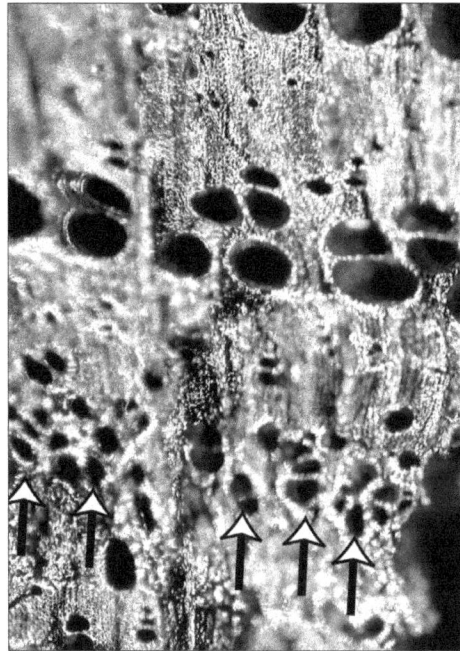

**Figure 6.** Narrow tree rings resulting from an earlier pollarding in an ash tree (*Fraxinus excelsior*) from an Early Neolithic layer of the Gardon cave (Ain, France) (cross section, enlargement: 100x).

**Figure 7.** Deciduous oak (*Quercus* f.c.) from a Late Neolithic layer of la Baume d'Oullins (Ardèche, France), showing several tree rings (about 34 for 1 cm) (cross section, enlargement: 50x).

**Figure 8.** Deciduous oak (*Quercus* f.c.) from a Late Neolithic layer of la Baume d'Oullins (Ardèche, France), showing a deformation on the multiseriate ray (cross section, enlargement: 50x).

## WOOD-ANATOMICAL EVIDENCE OF POLLARDING IN ARCHAEOLOGICAL CHARCOAL FRAGMENTS

As indicated by J.-E. Brochier (1996: 28) "From anthracological studies, we can remove only strong assumptions; other additional arguments are indispensable". If, for our part, we think that the over-representation of certain species in the anthracological results from sheepfold layers constitute a strong indication in favour of their use as fodder, we have now to test this assumption in the light of different technical approaches. One of the possibilities consists in the evidence of micro scars resulting from pruning. J.N. Haas and F.H. Schweingruber (1994) have shown that the effect of late summer pollarding on a broadleaved tree (ash tree for example) can be identified by a detailed analysis of the wood-anatomy. To demonstrate their model, cores were taken, with an increment borer, from trees (in Valais, Switzerland) with irregular crowns, indicating past interference with their branch development.

As ash pollarding usually takes place during the growing period (end of August, first half of September) every 4 or 6 years, the effects of pruning can be identified by the following two characteristics (HAAS AND SCHWEINGRUBER 1994: 37-38):
- "The first is that the ground meristem cells fail to thicken in the pollarding years. The tree rings remain narrow in the following year because the assimilates are used to build up the new crown...." (FIG. 4)
- "Buds for new shoots are formed in the same year the pollarding occured. This late summer blossom is associated with the formation of a second ring of earlywood pores in the latewood. Because these non seasonal earlywood pores are smaller than the normal earlywood pores they offer a good indication of pollarding."

As the authors write, it is "By the combination of both characteristic described above that it is possible to conclude that pollarding for leaf-fodder took place" (HAAS AND SCHWEINGRUBER 1994: 38).

To observe this pattern on archaeological burnt wood, we selected charcoal fragments of ash and deciduous oak, which presented some peculiarity from the sheepfold levels of the Gardon cave (Ain) and La Baume d'Oullins (Ardèche).

The Gardon cave, situated in the south of the Jura mountains, offers an exceptional stratigraphy which extends from the Mesolithic to the Iron Age. During Early and Middle Neolithic, the cave was visited by groups of shepherds and was used as sheepfold (VORUZ ET AL. 2004, SORDOILLET AND VORUZ 2002). At the time of the observation of the charcoal fragments, we tried systematically to disclose possible marks of pruning in the anatomical observation. Two examples from two different fragments of ash are presented here. They appear to us to be able to correspond to the observations realised by N. Haas and F. Schweingruber.

The first sample is an ash from a layer attributed to the Mid Neolithic. One can see in the cross section (FIG. 5) a row of small earlywood-like vessels in the late wood. This observation can be compared to the pattern proposed by Haas and Schweingruber (FIG. 4).

The second sample (FIG. 6) come from an Early Neolithic layer. The cross section shows an area of narrow tree rings quite comparable to those made on actual pollarded trees as the results of earlier pollarding.

The archaeological excavations of sheepfold layers discovered in La Baume d'Oullins cave provided numerous charcoal fragments from several layers attributed from the Early Neolithic Cardial to the Final Neolithic (ROUDIL 1987). Throughout the sequence, deciduous oak was harvested. Most of them presented a very particular anatomy with very tight rings (between 34 and 50 rings per cm) (FIG. 7), probably linked to an original management of the oak grove. Moreover, some charcoal fragments shows a marked deformation in the multiseriate ray which could be due to this mode of management.

The Figures 7 and 8 present a deciduous oak fragment, from a layer attributed to the Late Neolithic, showing very tight rings. Can the deformation observed in the multiseriate ray be in relation with the management of the tree?

Finally, this short paper aims at focusing the attention to the importance of different marks that can be observed on charcoal fragments. In the future, the development of anatomical approaches will allow the elaboration of new hypotheses concerning the management system for woodlands. This is a preliminary presentation, announcing a future research which should be developed in the next few years and whose purpose will be to look for, systematically, the consequences of pruning of trees by the anatomical study of archaeological charcoal fragments.

REFERENCES

AKERET Ö., JACOMET S., 1997.- Analysis of plant macrofossils in goat/sheep faeces from the Neolithic lake shore settlement of Horgen Scheller - an indication of prehistoric transhumance?, *Vegetation History and Archaeobotany,* 6: 235-239.

AKERET Ö., RENTZEL P., 2001.- Micromorphology and Plant Macrofossil Analysis of Cattle Dung from the Neolithic Lake Shore Settlement of Arbon Bleiche 3, *Geoarchaeology: An International Journal,* 16 (6): 687-700.

AKERET Ö., HAAS J.N., LEUZINGER U., JACOMET S., 1999.- Plant macrofossils and pollen in goat/sheep faeces from the Neolithic lake-shore settlement Arbon Bleiche 3, Switzerland, *The Holocene,* 9 (2): 175-182.

ARGANT J., HEINZ C., BROCHIER J.-L., 1991.- Pollens, charbons de bois et sédiments: l'action humaine et la végétation le cas de la grotte d'Antonnaire (Montmaur-en-Diois, Drôme), *Revue d'Archéométrie,* 15: 29-40.

AUSTAD I. 1988.- Tree Pollarding in Western Norway, *in:* H. H., Birks, H. J. B. Birks, P. E. Kaland and D. Moe (eds.), *The Cultural Landscape - Past, Present and Future,* New-York, Cambridge University Press: 11-29.

BECHMANN R., 1984.- *Des arbres et des hommes. La forêt au Moyen Age.* Paris, Editions Flammarion, 385 p.

BINTZ 1981.- *Les grottes de Choranche (Isère) - rapport de fouilles et études préliminaires,* Rapport dactylographié, Université de Grenoble, 29 p.

BREICHER H., CHABAL L., LECUYER N., SCHNEIDER L., 2002.- Artisanat potier et exploitation du bois dans les chênaies du nord de Montpellier au XIIIᵉ s. (Hérault, Argelliers, Mas-Viel), *Archéologie du Midi Médiéval,* 20: 57-106.

BROCHIER J.-E., 1996.- Feuilles ou fumiers? Observations sur le rôle des poussières sphérolitiques dans l'interprétation des dépôts archéologiques holocènes, *Anthropozoologica,* 24: 19-30.

BROCHIER J.-E., CLAUSTRE F., HEINZ C., 1998.- Environmental impact of Neolithic and Bronze Age farming in the eastern Pyrenees forelands, based on multidisciplinary investigations at La Caune de Bélesta (Bélesta Cave), near Perpignan, France, *Vegetation History and Archaeobotany,* 7: 1-9.

CHABAL L., 1997.- *Forêts et sociétés en Languedoc (Néolithique final, Antiquité tardive). L'anthracologie, méthode et paléoécologie,* Paris, Editions de la Maison des Sciences de l'Homme, 189 p. (Documents d'Archéologie Française, 63).

DUFRAISSE A., 2002.- *Les habitats littoraux néolithiques des lacs de Chalain et Clairvaux (Jura, France): collecte du bois de feu, gestion de l'espace forestier et impact sur le couvert arboréen entre 3700 et 2500 av. J.-C. Analyses anthracologiques,* Doctorat, Université de Franche-Comté, 349 p.

DURAND 1992.- Dynamique biogéographique des boisements forestiers en Languedoc durant le Moyen Age: l'impact de l'an mil, *in:* J.-L. Vernet (ed.), *Les Charbons de bois, les anciens écosystèmes et le rôle de l'homme (Montpellier, décembre 1991),* Paris, Editions Bulletin de la Société Botanique Française: 627-636. (Actualités Botaniques 1992-2/3/4).

DURAND-TULLOU A., 1972.- Rôle des végétaux dans la vie de l'homme au temps de la civilisation traditionnelle (étude ethnobotanique sur le Causse de Blandas, Gard), *Journal d'agriculture tropicale et de botanique appliquée,* XIX, 6-7: 222-246.

HAAS J.N., SCHWEINGRUBER F.H., 1994.- Wood-anatomical evidence of pollarding in ash stems from the Valais, Switzerland, *Dendrochronologia,* 11: 35-43.

HAAS J.N., KARG S., RASMUSSEN P., 1998.- Beech leaves and twigs used as winter fodder: Examples from Historic and Prehistoric Times, *Environmental Archaeology,* 1: 81-86.

HALSTEAD P., 1998.- Ask the fellows who lop the hay: leaf-fodder in the mountains of northwest Greece, *Rural History,* 9: 211–234.

HAEGGSTROM C.A., 1992.- Wooded meadows and the use of deciduous trees for fodder, fuel, carpentry and building purposes, *in:* J.-P. Métailié (ed.), *Proto-industries et histoire des forêts (octobre 1990, Toulouse),* Toulouse, GDR ISARD C.N.R.S.: 151-162.

LUDEMANN T., 2002.- Anthracology and forest sites- the contribution of charcoal analysis to our knowledge of natural forest vegetation in south-west Germany, *in:* S. Thiébault (ed.), *Charcoal analysis, methodological approaches, palaeoecological studies and wood uses. Proceeding of the second international meeting of anthracology (Paris, september 2000),* Oxford, BAR Publishing: 209-218 (BAR International Series 1063).

NELLE O., 2002.- Charcoal burning remains and forest stand structure - Examples from the Black Forest (south-west Germany) and the Bavarian Forest (south-east Germany), *in:* S. Thiébault (ed.), *Charcoal analysis, methodological approaches, palaeoecological studies and wood uses. Proceeding of the second international meeting of anthracology (Paris, september 2000),* Oxford, BAR Publishing: 201-208 (BAR International Series 1063).

NISBET R., 1997.- Arene Candide: charcoal remains and prehistoric woodland use, *in:* Maggi R., Starnini E. and Voytek B. (eds.), *Arene Candide: a functionnal and environmental assesment of the holocene sequence - excavations Bernabo Brea- Cardini, 1940-1950.* Roma, Il Calamo, vol. V: 103-112 (Memorie dell'Istituto Italiano di Paleontologia Umana).

PIQUÉ R., BARCELO J.A., 2002.- Firewood management and vegetation changes: a statistical analysis of charcoal remains form Holocene sites in the north-esat Peninsula, *in:* S. Thiébault (ed.), *Charcoal analysis, methodological approaches, palaeoecological studies and wood uses. Proceeding of the second international meeting of anthracology (Paris, september 2000),* Oxford, BAR Publishing: 1-8 (BAR International Series 1063).

RASMUSSEN P., 1989.- Leaf Foddering of livestock in the Neolithic archaeobotanical evidence from Weier, Switzerland, *Journal of Danish Archaeology,* 8: 51-71.

RASMUSSEN P., 1993.- Analysis of Goat/Sheep Faeces from Egolzwil 3, Switzerland: Evidence for branch and twig foddering of livestock in the Neolithic, *Journal of Archaeological Science,* 20: 479-502.

ROUDIL J.-L., 1987.- Le gisement néolithique de la Baume d'Oullins, le Garn - Gard, *in:* J. Guilaine (ed.), *Premières communautés paysannes en Méditerranée occidentale (Montpellier, avril 1983),* Paris, Editions C.N.R.S.: 523-529.

SORDOILLET D., VORUZ J.-L., 2002.- Un nouvel enregistrement climatique dans un système karstique, la stratigraphie du Gardon, *in:* H. Richard and A. Vignot (eds.), *Equilibres et ruptures dans les écosystèmes depuis 20000 ans en Europe de l'Ouest, Actes du colloque international de Besançon septembre 2000,* Besançon, Presses Universitaires Franc-Comtoise: 91-106 (Annales Littéraires n°730, série Environnement, Sociétés et Archéologies n°3).

THIÉBAULT S., 1988.- *L'homme et le milieu végétal - analyse anthracologique de six gisements des Préalpes sud-occidentales aux Tardi et Postglaciaire,* Paris, Editions de la Maison des Sciences de l'Homme, 112 p. (Documents d'Archéologie Française, 15).

THIÉBAULT S., 1997.- Holocene vegetation and human relationships in central Provence area: charcoal analysis of Baume de Fontbrégoua (Var, France), *The Holocene,* 7 (3): 341-347.

THIÉBAULT S., 1999.- *Dynamique des paysages et intervention humaine du Tardiglaciaire à l'Holocène, de la Méditerranée aux Préalpes sud-occidentales - apport de l'analyse anthracologique,* Habilitation à Diriger des Recherches, Université de Paris I, 278 p.

THIÉBAULT S., 2001.- Anthracoanalyse des établissements néolithiques de la région liguro-provençale, *Bulletin de la Société Préhistorique Française,* 98 (3) : 399-409.

THIÉBAULT S., IN PRESS.- L'apport du fourrage d'arbre dans l'élevage depuis le Néolithique, Actes de la table-ronde de Penne 2004, *Anthropozoologica.*

VERNET J.-L. 1997.- *L'homme et la forêt méditerranéenne de la Préhistoire à nos jours,* Paris, Editions Errance, 248 p.

VORUZ J.-L., PERRIN T., SORDOILLET D. *ET AL.,* 2004.- La séquence néolithique de la grotte du Gardon (Ain), *Bulletin de la Société Préhistorique Française,* 101 (4): 827-866.

# EVIDENCE OF TRIMMED OAKS (*QUERCUS* SP.) IN NORTH WESTERN FRANCE DURING THE EARLY MIDDLE AGES (9TH-11TH CENTURIES A.D.)

VINCENT BERNARD, SYLVAIN RENAUDIN, DOMINIQUE MARGUERIE

UMR 6566 "Civilisations Atlantiques et Archéosciences"
Université de Rennes 1
Beaulieu F-35042 Rennes cedex (France)
vincent.bernard@univ-rennes1.fr
sylvain.renaudin@lycee-jean-queinnec.org
dominique.marguerie@univ-rennes1.fr

ABSTRACT: In North-western France, pollarding is still a living practice very linked to the management of a typical landscape, where hedgerow network or "bocage" constitute the main component. In Brittany, pollarding takes a particular form called "trimming" or "shredding" (Rackham, 1980). Trimming consists in cutting down regularly all of the branches of the trees, specially oak (*Quercus robur*) and chestnut trees (*Castanea sativa*). This study aims to specify the nature of the tree's reaction to this stress and adapt these event years for archaeological purposes to grasp the origin and the evolution of our existing landscape.The comparison of the tree-ring signal between living trees coming from five sites in eastern Brittany and archaeological timbers from four medieval sites in the Paris basin and the Armorican massif show very similar anatomical anomalies. This analysis gives a key to ensure the association of growth stress cycles observed on ancient woods to the practice of trimming. But the high frequency of archaeological timbers with clear signs of trimming asks the question of a possible existence of a "bocage" landscape during the early Middle Ages.
KEY WORDS: Trimming, Pollarding, Ring width, Wood anatomy, Dendrological signature, Event years, Archaeology, Brittany, France

RÉSUMÉ: L'émondage reste encore aujourd'hui une pratique vivace dans le Nord-Ouest de la France, où il contribue largement à l'entretien du bocage et de ses réseaux de haies. En Bretagne, l'arbre émondé prend une forme particulière que l'on appellera localement "ragosse", "émonde" ou "têtard". Cette pratique bocagère consiste à élaguer totalement et régulièrement les arbres, et plus particulièrement les chênes pédonculés (*Quercus robur*) et les châtaigniers (*Castanea sativa*).Cette étude a donc pour but de définir la nature des réactions d'un arbre à ce type de stress et d'adapter la signature de l'émondage au matériel archéologique, afin d'appréhender l'origine et l'évolution de notre paysage actuel.La comparaison entre le signal dendrologique d'arbres actuels issus de cinq sites de Haute-Bretagne et celui de bois anciens provenant de quatre sites médiévaux du Bassin parisien et du Massif armoricain montre de grandes similitudes dans les anomalies anatomiques observées. Cette analyse nous livre ainsi une clef pour relier en toute confiance des cycles de chutes de croissance enregistrées par des bois anciens à la pratique de l'émondage. Mais, la question d'un bocage dès le haut Moyen Age reste posée, sur la base d'assez nombreux éléments architecturaux issus manifestement d'arbres émondés.
MOTS-CLÉS: émondage, largeur de cerne, anatomie du bois, signature dendrologique, archéologie, Haute-Bretagne, France

ZUSAMMENFASSUNG: In Nordwestfrankreich besteht die Tradition des Schneitelns bis heute fort und trägt damit massgeblich zur Erhaltung des Landschaftstyps des "Bocage" mit seinen netzwerkartigen Hecken bei. In der Bretagne nimmt der geschneitelte Baum eine spezielle Gestalt an. Beim Schneiteln werden dem Baum, vor allem Stieleichen (*Quercus robur*) und Kastanien (*Castanea sativa*), in regelmässigen Zeitabständen sämtliche Äste abgeschnitten.Diese Studie hat das Ziel, die charakteristischen Reaktionen des Baumes auf diesen Stressfaktor zu definieren, damit auch an archäologischem Material erkannt werden kann, wenn ein Baum geschneitelt wurde. Dies trägt zum besseren Verständnis von Herkunft und Entwicklung unserer heutigen Landschaft bei.Die Baumringmuster von heute existierenden Bäumen (fünf Standorte in der Bretagne) und von mittelalterlichem Holz (Fundorte im Pariser Becken und dem Massif armoricain) weisen in ihren anatomischen Anomalien grosse Ähnlichkeit auf. Die Analyse lässt den Schluss zu, dass die am mittelalterlichen Holz beobachteten Wachstumseinbrüche auf die Praxis des Schneitelns zurückgeführt werden können. Der hohe Anteil an mittelalterlichem Holz mit Schneitelhinweisen lässt allerdings die Frage offen, ob es den "bocage" bereits im Hochmittelalter gegeben hat (*Translation Petra Zibulski*).
STICHWORTE: Schneiteln, Jahrringbreite, Holzanatomie, Weiserjahr, Archäologie, Bretagne, Frankreich

North-western France has today the largest area of hedgerow network landscape or "bocage" in the country (FIG. 1). The farmers' activities firstly consist of structuring features on the land, such as field patterns and field boundaries, including hedgerows. The "traditional" hedgerows are predominant in the landscapes of eastern Brittany that we are studying: the plantation of these hedgerows culminated in the 19th century. They include oak (*Quercus robur*), chestnut (*Castanea sativa*) and sometimes ash (*Fraxinus excelsior*) or beech (*Fagus sylvatica*), with a majority of pollarded trees.

## CURRENT TRIMMING PRESENTATION AND STUDY AREA

Trimming consists of cutting down all of the branches of a tree, from the bottom to the top, in order to gather its ligneous matter, mainly for heating purposes (LIZET AND RAVIGNAN 1987, RACKHAM 1980). In North-western France, this operation takes place in winter and at 7-12 year intervals.

Currently, there are two traditional and distinctive types of trimming in Brittany: the well-known "pollarded trees" (oaks, ashes and chestnut trees), which are 1.80 to 5 meters high, and the shredded trees (exclusively oaks) which are 4 to 6 meters high (RENAUDIN 1996).

The distribution area of the shredded trees is centred in the western zone of Brittany in the Rennes basin (FIG. 1).

Beyond the clear traces on the external morphology of the tree, the impact of pruning can be evidenced by ring-width series and wood anatomy. These morphological and anatomical criteria easily observable optically on ancient oak timbers, have allowed us to specifically select material on archaeological and historical sites.

Few authors have been able to show the ability of the ligneous plant to react to harsh exogenous disturbances of human origin (GUIBAL 1988, CHRISTENSEN AND RASMUSSEN 1991, HAAS AND SCHWEINGRUBER 1994).

The disturbance of the ligneous productivity in the radial growth of the wood analysed here concerns the trimming of deciduous oaks.

This study aims to (i) specify the nature of the tree's reaction to this stress by bringing the dendrological "trimming signature" to light, (ii) adapt these event years for archaeological purposes to grasp the origin and the evolution of our existing landscape where for ages hedges have been one of the main wood supplies.

## MATERIAL AND METHODS

The living trees are *Quercus robur L.* which presents a wood with porous zones and come from five sites in eastern Brittany (FIG. 1). Some sections have been sampled at 1.20 m from the ground on 16 shredded trees, and a core has been sampled on untrimmed trees on each site, in order to compare the two populations. The years of the last trimmings are known for each sample thanks to an inquiry with the farmers.

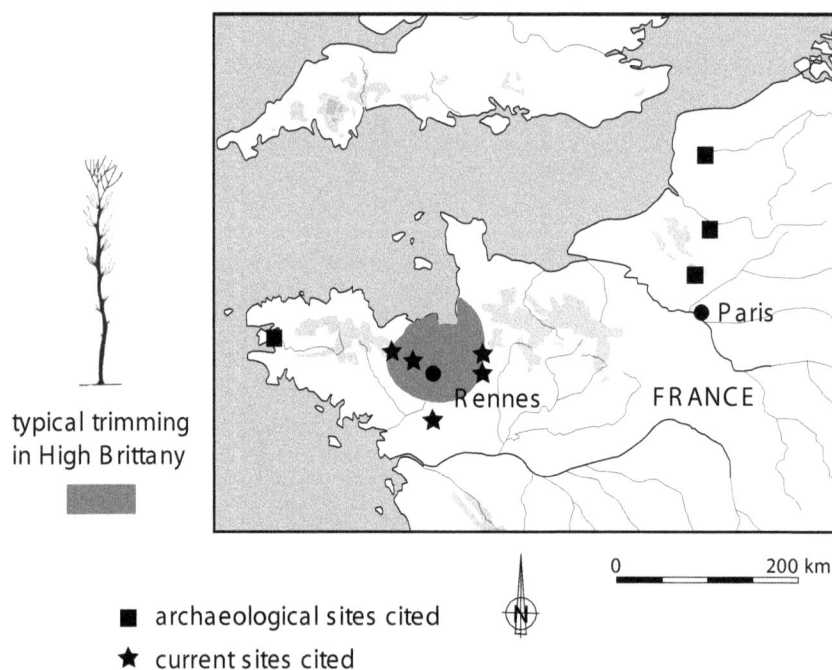

typical trimming in High Brittany

■ archaeological sites cited
★ current sites cited

**Figure 1.** Study area and location of the study sites.

This has enabled us to measure the ring widths corresponding to 47 trimming cycles. The anatomy of 6 out of 47 cycles has been analysed by means of a microtome.

As for historical and archaeological oak timbers (*Quercus sp.*), they come from a larger area including the Armorican massif and the Paris basin (FIG. 1). The main part of this corpus comes from the frameworks of medieval churches and manors, but the more ancient and the more original elements were discovered in archaeological layers dating from the Carolingian period (9th-10th centuries A.D.). These four sites belong to building phases dated by dendrochronological analysis as having occurred between 810 and 950 A.D. A total of 300 architectural timbers (posts, planks, timbers, etc.) were measured, but less than 20 presented similar growth stress. In fact, these 20 samples are not dated directly by tree-ring analysis, only by indubitable association with well dated woodworks. F. Guibal clearly showed that this sort of material is useless for dendrochronology, since the climatic signal is totally disturbed (GUIBAL 1988).

The tree-ring widths and anatomy inside the trimming disturbances were measured in the dendrochronological laboratory of the Centre d'études nordiques of the Laval university of Quebec, Canada thanks to the N.I.H. Image software (American public ownership). Furthermore, the tree-ring measurements have been carried out in France using a digital positioning dendrochronological table (Lintab Rinntech, accuracy: 0.01 mm).

Several dendrological variables are taken into account:
- ring width;
- earlywood and latewood width;
- vessel surface;
- number of vessels;
- vessel position in the ring.

By comparing the undisturbed rings and the disturbed ones following to the trimming, all the data, if they are pertinent, will reveal the dendrological signature of the trimming operation.

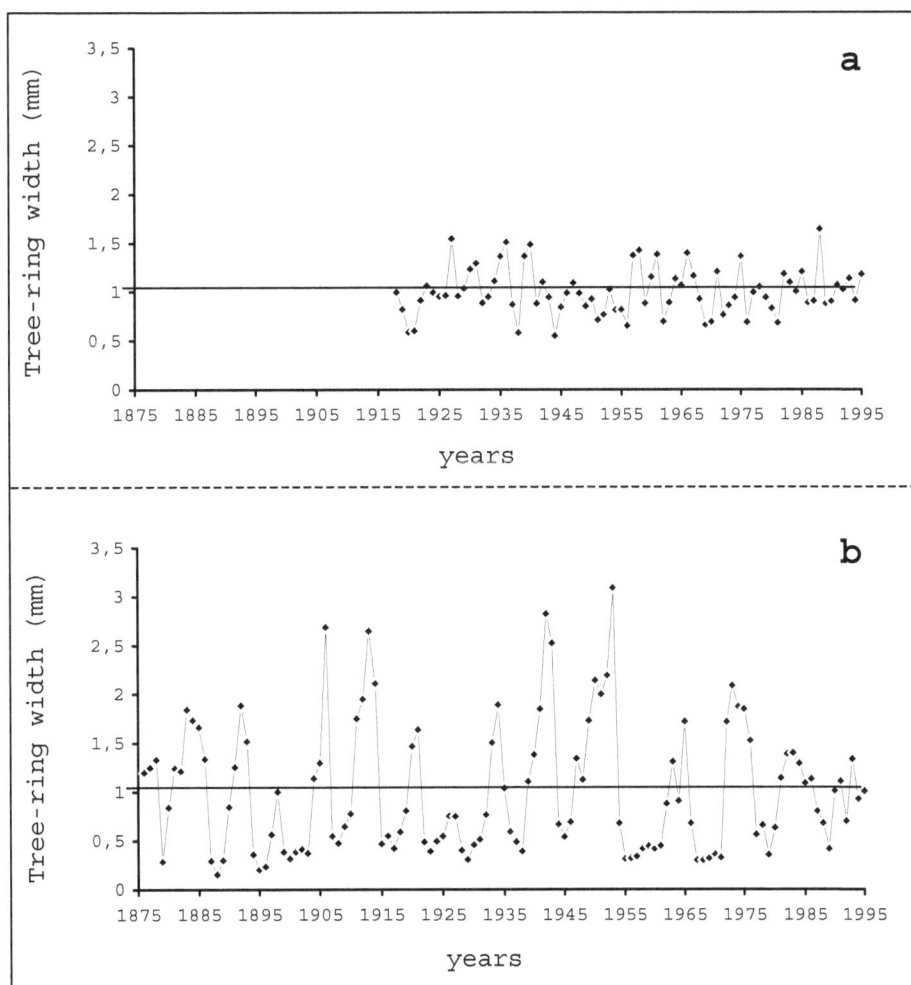

**Figure 2.** Dendrochronological curves of one untrimmed oak (a) and a trimmed current oak (b) (Saint-Germain-de-Coulamer).

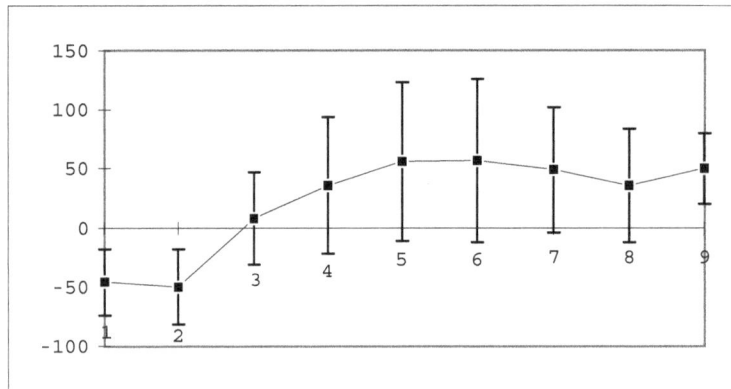

**Figure 3.** Ring width variations several years after the trimming.

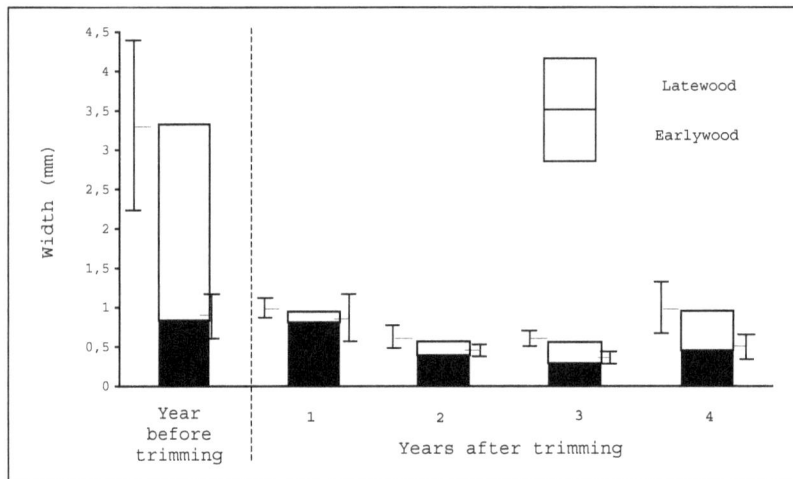

**Figure 4.** Widths variations of earlywood and latewood before and after trimming (n = 6 cycles).

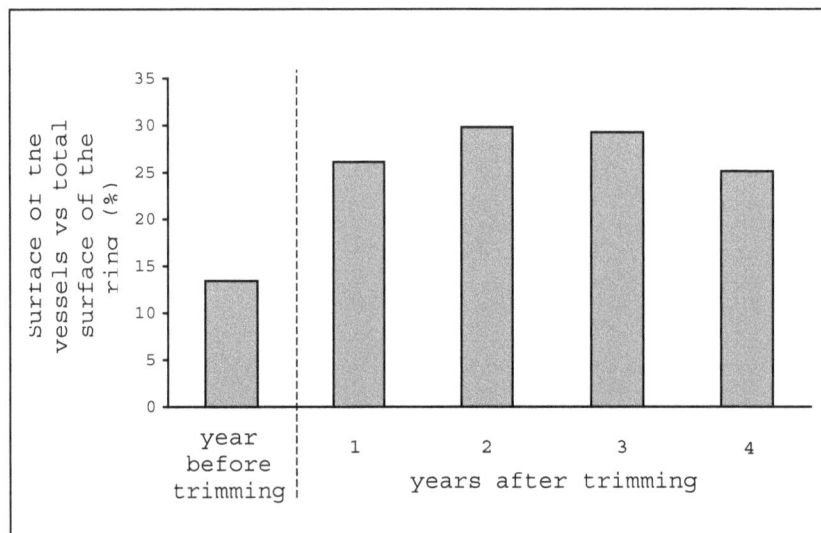

**Figure 5.** Mean of porosity of the rings (%) before and after the trimming (n = 6 cycles).

## RESULTS

### Current trees

*Ring width*

The ring width measurements show similarities between trimmed trees: the ring widths of the trimmed trees are affected by abrupt growth changes during the 3 to 4 years following the trimming. These abrupt growth changes ("trimming valleys") appear regularly, every 7 to 12 years, on the curves representing all the trimmed trees of the different sites. These measurements show similarities between trimmed trees and differences between trimmed and untrimmed trees. The untrimmed trees do not show such growth variations (FIG. 2).

The measurements indicate negative variation rates of ring width during the first 2 years (year 0: -46%, y+1: -50%) and positive variation rates the following years (y+2: +8%). The second year following the trimming is the year when the variation rate is the smallest. The biggest variation of the values after the third year shows that the effect of trimming on the ring widths softens (FIG. 3).

*Ring anatomy*

The anatomical analysis of the wood of the trimmed trees shows characteristic anomalies.

The earlywood width is not altered for the first year, which follows the trimming ("year 0"). Its proportion within the ring strongly increases because of the extremely low proportion of latewood (from 15-35% within the ring before the disturbance to up to 90% within the ring after the disturbance). Furthermore, during the "year+1" and the "year+2", the widths of the earlywood and of the whole ring are the smallest, despite the latewood, the width of which increases in small proportions in comparison with the first year. After 3 to 4 years, the early wood width increases at the same time as the latewood width. Afterwards, the earlywood proportion tends to decrease in comparison with that of the latewood and as the ring width increases (FIG. 4).

The average surface area of the vessels in the early wood is also significant: it is clearly inferior in year+1, +2 and +3 whereas it is not altered in year 0 (FIG. 5).

### Historical and archaeological timbers

Compared to the investigation of recent material, dendrological analyses were carried out by using the same techniques on a series of medieval timbers from four sites, where tree-ring sequences show several abrupt growth changes. These four sites belong to Carolingian building phases dated by dendrochronological analysis to between 810 and 950 A.D, Belle-Eglise, Saleux and Douai in the Paris basin and Landevennec in Brittany (BERNARD 1998). A total of 300 architectural woods (posts, planks, timbers...) were measured, but less than 20 presented similar growth stress: abrupt growth changes appear every 7 to 9 years and during the last 3 years. In spite of the small number analysed, these samples are sufficiently significant by their tree-ring series and their anatomy to be associated to the practice of trimming (BERNARD 1997). Anatomically, the two different types of trimming, pollarding and shredding, cannot be distinguished. But, in this case, we can consider that shredded trees (and not pollarded trees, because of their tall height and their typical appearance) were being maintained during High Middle Ages.

## DISCUSSION

Among the signals that enable us to say that the studied trees have been affected by a disturbance, the dendrological "trimming signature" is characteristic on the ring widths and in the ring anatomy. Thanks to the regularity of the phenomenon, the absence of known infestations, the observation of these growth anomalies on each trimmed tree studied and the knowledge of the dates of the last trimmings, we can affirm these growth anomalies are due to trimmings.

Besides bringing the dendrological signature of the oak trimming to light, this study is important with regard to archaeological research. When comparing samples of preserved ligneous traces (charcoals, woods full of water, dry woods) with this signature, then we can decide, without risks, whether the wood comes from a trimmed tree or not. In the case of small charcoals, the observation of several periods of trimming is impossible or rare. If a series of minimum three periods of trimming is not preserved, it is impossible to attest that the charcoal comes from a trimmed tree (MARGUERIE AND HUNOT IN SUBMISSION).

Archaeological samples coming from sites analysed within this work, have enabled us to determine if the trees have been affected by either trimming as it is practised today, or by a disturbance similar to trimming which also produces a loss of foliage (RASMUSSEN 1988).

The high frequency of archaeological timbers with clear signs of trimming on a given site could mean the possible presence of a bocage or hedgerow network landscape. However, can we necessarily correlate "pollarded trees" and "bocage"? Future studies need to coordinate dendrochronological and palynological analysis, which, if shown to give encouraging results, could become an interesting spatio-temporal tool in the research of the setting up and the global evolution of the bocage in these regions (BAUDRY AND JOUIN 2003, MARGUERIE *ET AL.*, 2003).

ACKNOWLEDGEMENTS

This present work is the result of a scientific collaboration between the "Centre d'Etudes Nordiques" (CEN, University of Laval, Canada) and the laboratory "Civilisations Atlantiques et Archéosciences" (CNRS/University of Rennes 1, France). The image analysis has been operated thanks to a measuring system developed by Luc Cournoyer (CEN). The supervision of the English text was made by John Vaughan.

REFERENCES

BAUDRY J., JOUIN A., 2003.- *De la haie aux bocages, organisation, dynamique et gestion*, Paris, Editions de l'I.N.R.A., 435 p. (Espaces ruraux).

BERNARD V., 1998.- *L'homme, le bois et la forêt en France du nord du Mésolithique au Haut Moyen-Age*, Oxford, BAR Publishing, 190 p. (BAR International Series 733).

CHRISTENSEN K., RASMUSSEN P., 1991.- Styning af traeer (pollarding of trees), *Eksperimentel Arkaeologi:* 24-30.

GUIBAL F., 1988.- Aspects de la dendrochronologie des habitations seigneuriales de Bretagne, *in:* Hackens, A.V. Munaut and C. Till (eds), *Wood and Archaeology (Bois et Archéologie). First European Conference (Louvain-la-Neuve, October 2nd-3rd 1987)*, Strasbourg, PACT: 85-97.

HAAS J.-N., SCHWEINGRUBER F.H., 1994.- Wood anatomical evidence of pollarding in ash stems from the Valais, Switzerland, *Dendrochronologia*, 11: 35-43.

LIZET B., DE RAVIGNAN F., 1987.- *Comprendre un paysage. Guide pratique de recherche. Espaces ruraux*, Paris, Editions de l'I.N.R.A., 150 p.

MARGUERIE D., HUNOT J.-Y., IN SUBMISSION.- Charcoal analysis and dendrology: data from archaeological sites in north-western France, *Journal of Archaeological Science*.

MARGUERIE D., ANTOINE A., THENAIL C., *ET AL.*, 2003.- Bocages armoricains et sociétés, genèse, évolution, interaction, *in:* T. Muxart, F.D. Vivien, B. Villalba and J. Burnouf (eds.), *Des milieux et des hommes: fragments d'histoires croisées*, Paris, Editions Elsevier: 115-131.

RACKHAM O., 1980.- *Ancient woodland. Its history, vegetation and uses in England*, London, E. Arnold, 376 p.

RASMUSSEN P., 1988.- Pollarding of trees in the Neolithic: often presumed, difficult to prove, *in:* D.E. Robinson (ed.), *Experimental and reconstruction in environmental archaeology, AEA Symposia 9 (Roskilde, DK)*, Oxford, Oxbow Books, 9: 38-45.

RENAUDIN S., 1996.- *Les émondes de Haute-Bretagne: Etude dendrologique du chêne et perspective archéologiques*, DEA, Université de Nantes, 86 p.

# PARTICIPANTS LIST

NIELS BLEICHER
Universität Göttingen, Germany

YOLANDA CARRION
Departamento de Prehistoria y Arqueología
Universitat de València, Spain

FREDDY DAMBLON
Royal Belgian Institute of Natural Sciences,
Belgium

CLAIRE DELHON
UMR 7041 CNRS Archéologies et Sciences de
l'Antiquité, Université Nanterre Paris X, France

MARY DILLON
Palaeoenvironmental Research Unit
Department of Botany, National University of Ireland,
Galway, Ireland

ALEXA DUFRAISSE
Laboratoire de Chrono-écologie
UMR 6565 CNRS, Université de Franche-Comté, France

STEFANIE JACOMET
Institut für Prähistorische und Naturwissenschaftliche
Archäologie, Universät Basel, Switzerland

ANUSCHKA JAHNKE
Universität Freiburg, Institut für Biologie II/Geobotanik,
Freiburg, Germany

CATHERINE LAVIER
Laboratoire de Chrono-écologie,
UMR 6565 CNRS, Université de Franche-Comté, France

THOMAS LUDEMANN
Forest Research Institute of Baden-Württemberg
Institute of Biology II, Freiburg i.B., Germany

DOMINIQUE MARGUERIE
Civilisations atlantiques et archéosciences,
UMR 6566 CNRS, Université de Rennes 1, France

DANIÈLE MARTINOLI
Institut für Prähistorische und Naturwissenschaftliche
Archäologie, Universät Basel, Switzerland

JUTTA MEURERS-BALKE
Labor für Archäobotanik
Institut für Ur- und Frühgeschichte
Universität zu Köln, Germany

DENIS MORIN
UMR 5608 Unité Toulousaine d'Archéologie et
d'Histoire, Toulouse, France

VANESSA PY
Laboratoire d'Archéologie Médiévale Méditerranéenne,
UMR 6572 CNRS, Université de Provence
Aix-en-Provence, France

WERNER SCHOCH
Laboratory for Quaternary Wood Research
Langnau, Switzerland

FRITZ H. SCHWEINGRUBER
Swiss Federal Institute of Forestry Research
Birmensdorf, Switzerland

INGELISE STUIJTS
The Discovery Programme
Dublin, Ireland.

URSULA TEGTMEIER
Labor für Archäobotanik
Institut für Ur- und Frühgeschichte
Universität zu Köln, Germany

STÉPHANIE THIÉBAULT
UMR 7041 CNRS Archéologies et Sciences de
l'Antiquité, Université Nanterre Paris X, France

JEAN-LOUIS VERNET
Centre de Bioarchéologie et d'Ecologie,
UMR 5059 CNRS et Université de Montpellier,
France

STEVEN WARD
Oxford Long Term Ecology Laboratory
University of Oxford, United Kingdom

PETRA ZIBULSKI
Institut für Prähistorische und Naturwissenschaftliche
Archäologie, Universät Basel, Switzerland

www.ingramcontent.com/pod-product-compliance
Lightning Source LLC
Chambersburg PA
CBHW061008030426
42334CB00033B/3403